Crush

MY YEAR *AS AN* APPRENTICE WINEMAKER

NICHOLAS O'CONNELL

Potomac Books

An imprint of the University of Nebraska Press

 Potomac Books is an imprint of the University of Nebraska Press.
Manufactured in the United States of America.

For customers in the EU with safety/GPSR concerns, contact:
gpsr@mare-nostrum.co.uk
Mare Nostrum Group BV
Mauritskade 21D
1091 GC Amsterdam
The Netherlands

Library of Congress Control Number: 2024048524

Designed and set in Bulmer MT Std by Lacey Losh.

For Dan McCarthy

CONTENTS

ILLUSTRATIONS

Following page 114

ACKNOWLEDGMENTS

Writing *Crush* was a labor of love, but it would have been impossible without the help of many people. I'm especially indebted to the winemakers, viticulturalists, and others who gave so generously of their time, attention, and knowledge: Robert Mondavi, Warren Winiarski, Michael Silacci, Chris Tomkiewicz, Gordon Walker, Juan Martinez, Chris Howell, Ashley Anderson-Bennett, François Bugué, Marcus Notaro, Nicki Pruss, Kirk Grace, Jim Holmes, Richard Holmes, Kade Casciato, Bob Betz, Louis Skinner, Jonathan Villaseñor, Chris Upchurch, Chris Peterson, Craig Nickel, Ross Mickel, David Lake, Scott Willams, JJ Williams, Tim Hightower, Rob Griffin, Travis Maple, John Bookwalter, Brian Mackey, Norm McKibben, Jean-François Pellet, Gary Figgins, Chris Figgins, Rick Small, Haydn Mouat, Paul Portteus, Mike and Mikey Etzel, Omar Perea, Tony Soter, Chris Fladwood, Hallie Whyte, Ken Wright, Carson Wright, J'Aime du Mauriée, Seth Miller, Ivory McLaughlin, Joylin Kent, Daniel Ravier, and Lulu Peyraud.

Publishing a book is always a herculean task, but having a great team makes it possible. I'm especially grateful to University of Nebraska Press acquisitions editor Taylor Gilreath, project editor Ryan Masteller, copy editor Vicki Chamlee, lead senior designer Lindsey Welch, and marketing and sales manager Rosemary Sekora for their warmth and professionalism in accepting, editing, designing, and promoting the book.

I'd also like to acknowledge the magazine editors whose assignments helped inspire *Crush*, especially Paul Frichtl and Michele Dill of *Alaska Airlines Magazine* and Allen Cox of *Northwest Travel Magazine*. I'd like to thank Lynda Eller, senior director of communications and corporate affairs of Chateau Ste. Michelle; Dan McCarthy of McCarthy & Schiering Wine Merchants; and oenologist Erica Orr for their support and technical advice.

I owe a particular debt to our Les Copains group, especially Tom Remmers, my wine-making partner of thirty years; Bruce McLachlin; Gregory

Heller; Theresa Jurotich; Jen Harris and her dog Banjo; William Ayre; Jerry Casson; Abbey Linh Bennet; Steve Rottler; Scott Driscoll; Angus Duggan; Chris Olsen; Ron Weglin; Anne Weglin; Cathy Rouleau; Tad Lewis; Victoria Albanese; Ehben Eliot; David George; Mark Jenkins; Bob Berg; Andrew Rice; Mike Medberry; Steven Albright; Paula Cipolla; and others.

Finally, I'd like to thank my family for indulging my wine obsession, especially my wife, Lisa Sowder, who gave up garage space for wine storage, served piles of pasta at our annual bottling party, and survived an infestation of fruit flies; my sons Daniel and Nick Jr.; and my daughter, Marie, for accompanying me on my travel, food, and wine adventures.

Crush

PART 1

Crush

1

Garagiste

Winemakers call it the Long Haul. It's a three-hour drive from Seattle to eastern Washington, where the state's best wine grapes grow. The state's wineries are largely located on the wet west side, while the vineyards are located on the dry eastern side. Every fall, winemakers make the 180-mile run.

My wine-making partner, or *garagiste* (garage winemaker), Tom Remmers lends me his ancient Ford F250 diesel, nicknamed "the Beast," to pick up 1,200 pounds of Cabernet grapes from Ciel du Cheval Vineyard on Red Mountain—one of the finest red wine grape regions in Washington State and arguably the planet. I'd prefer to take my van, which is more reliable, but Tom argues the truck is the better choice for transporting that many grapes.

While I fetch the grapes, Tom prepares for the crush, setting up the big plastic fermenting tanks, cleaning out the stemmer/crusher, and assembling the crew. Timing is critical, with little margin for error, as the grapes need to be crushed as soon as possible after picking. Making the three-hour run should be a seemingly simple task, but nothing about wine making is simple.

The Beast shudders and shakes as I pull out of the driveway. Climbing a modest incline, it shimmies, backfires, and belches a cloud of black smoke. After merging onto I-90, I lean back into the Naugahyde bench seat. I've been making this run for fifteen years now, but this is the first time with the Beast. I was reluctant to take it, given its 250,000 miles on the odometer, but Tom encouraged me. "You'll love it," he said. "You can have them dump the grapes in the back. You don't need to put them into buckets."

This trip is the second of three runs to pick up grapes to make wine for our wine co-op, Les Copains ("the friends" in French). While I love the romance of wine making, I wonder, as do many aspiring amateurs, if I could take it to the next level. I've worked for years to gain access to the best fruit, with the famous boutique wineries gobbling up all but a fraction of it. A few years ago, Ciel du Cheval Vineyards (translation of nearby Horse Heaven Hills) agreed to sell to us as I write about wine and so am considered part of the industry. I don't want to do anything to jeopardize getting fruit from one of the best vineyards in the state, a critical step in making outstanding wine. We've won first place awards at amateur competitions. I should be satisfied, but there seems so much more to learn.

After crossing Snoqualmie Pass, the low point in the Cascade Range that vertically bisects the state, I turn south onto I-82, which rises steeply toward Umtanum Ridge. The truck grumbles as the grade increases. I keep my foot on the gas, willing it up the hill. My coffee cup dances along the top of the dashboard. A burnt smell seeps out of the heater. I keep it floored, praying the heap will make it over the top. Black, acrid smoke billows out of the dashboard vent. I open the window and stick my head outside to avoid asphyxiation. I ease up on the pedal, but the smoke keeps coming. Finally, the truck clears the top of the ridge.

I pull off to the side and let the Beast cool off. Leaning against the side panel, I take a drink of water, resolving *never* to drive this piece of shit again! But I've got to get the Beast to Ciel. The grapes are ready. If you pick too early, the grapes taste green and vegetal. If you pick too late, they have too much sugar and produce a strong, alcoholic wine. You have to pick them right on time. Our crew of a half-dozen Les Copains members has already committed to helping with the crush. Turning back is not an option.

With the smoke cleared, I fire up the truck and keep going. The road traverses the undulating brown hills and scabrock country outside Yakima. I keep my foot on the gas, coaxing the Beast up the hills and making up time on the downhill, hitting sixty.

My appointment is at 1:00 p.m. with Ryan Johnson, the manager at Ciel du Cheval Vineyard, located outside of the Tri-Cities. Ciel has produced several hundred-point scores from renowned wine critic Robert Parker Jr., and dozens of others were rated in the nineties over the last decade. Beyond the scores, I consider this my favorite site, with bold, powerful red wines that display balance and elegance, similar to some of the best French wines but with a richer, rounder new-world taste.

I glance at my watch: 12:30 p.m. Right on track. I'm fifteen minutes away from Ciel, the Beast humming along the straightaway.

KABOOM! The truck bucks violently and veers to the right, heading for the ditch. I whip the wheel to the left, overcompensate, and swerve into the passing lane, narrowly missing a truck. I swerve back to the right, tires squealing on the asphalt. The Beast bobs and weaves like a rodeo bull, trying to throw me. Slamming on the brakes, I fight to regain control. Finally, I wrestle it to a stop on the shoulder of the highway.

Breathing rapidly, I look around. What was that? A stray tank round from the nearby U.S. Army's Yakima Training Center? The cab doesn't seem to have any physical damage, other than the dregs of my coffee spilled on the floor. I breathe deeply to slow the pounding of my heart. Taking a look to make sure no cars are heading my way, I open the driver's door and get out. I walk around to the right side of the truck. The right front tire looks as if someone carved it up with a chainsaw. There are scorch marks behind the truck on the asphalt. The smell of burned rubber fills the air.

At the back of the truck, I spot a spare tire underneath the truck bed. The bolts securing it are rusted in place. Damn! I call Ryan on my cell phone.

"I'm going to be a little late," I say, trying to sound calm. "I've got a flat tire. Is there any place around here that can fix it?"

"There's a Les Schwab in Benton City," he says. "Can you get there?"

"I'll try," I reply, hoping he's not judging me a bumbling amateur.

I start up the truck again. The diesel engine coughs and sputters to life. Slowly, I push down on the gas pedal. The Beast lurches forward, making a horrible metallic grating sound as the shredded tire grinds into the asphalt. I keep going, the tire smoking and squealing, a ragged cloud of burned rubber trailing behind. I keep the speedometer at five miles an hour and pray the truck will make it to the Les Schwab station.

Turning off the highway, I head for downtown Benton City. There's one main street with gas stations and feed stores. People stare as the Beast belches smoke and limps along like a wounded water buffalo. I'm providing the afternoon's entertainment.

I spot the Les Schwab sign and turn into the lot. A young man greets me and frowns at the tire. "Looks like you need a new tire," he says enthusiastically.

While he replaces the tire, I walk down to a store to buy a diet soda. I shake my head about the flat; the rubber was probably so old it just fell apart. Would the other tires do the same? By the time I return, he has replaced the old tire. I pay and thank him. Then I rev up the Beast and get back on the highway.

The road to Red Mountain rises above Benton City, passing fields of sagebrush and rusted-out tractors. It looks more like farm country than a world-class vineyard site, but so did Napa Valley thirty years ago. "Welcome to Red Mountain," proclaims a stone and wood sign of a vaguely old-world design, the only indication I'm approaching a famous viticultural area. Low brown hills rise in the distance, dotted with sagebrush and covered with shattered volcanic rock. Beneath them appear the bright green geometric patterns of vineyards. There are no restaurants, hotels, or wine trains on Red Mountain; the area is as pure and abstract in its undulating beauty as a landscape in Tuscany.

It is just after 2:00 p.m. when I turn right on a gravel road lined with poplars. I'm an hour late, but at least the truck hasn't broken down again. I pass a Ciel du Cheval sign, a rusted-out truck, and then strict rows of vines with carefully tended clusters of Cabernet, Merlot, and Syrah grapes hanging beneath them, reflecting the perfectionism of owner Jim Holmes and his manager Ryan Johnson. The astonishing Ciel fruit reveals the power and elegance of the Red Mountain region. It's a wine lover's idea of heaven.

Rising up from the vines is not a chateau but a large wooden outbuilding, a kind of glorified tractor shed, surrounded by workers. Forklifts ferry bins of grapes to waiting trucks, loading them up for the long haul to the west side. The bins are stenciled with the names of the vineyard's customers: Quilceda Creek, DeLille, Andrew Will, McCrea Cellars, Cadence, Betz Family, Fidélitas, Mark Ryan—a who's who of Washington wine. The place hums with order and purpose. The harvest is in full swing.

Ryan comes out as I arrive at the front of the building. "Sorry I'm late," I say, getting out of the truck. I shake hands with Ryan, an energetic guy in his thirties with a freckled face, a reddish goatee, and a spring in his step. He wears jeans, a white shirt, and a floppy sun hat. During the harvest, he serves as a kind of air traffic controller, overseeing the picking and distribution of the grapes. This job is challenging, as grapes must be picked on time to the specifications of thirty or so winemakers, who are a notoriously fussy and individualistic lot. Johnson deftly handles the job, loading everything from huge semis to trucks like my own.

"Don't worry about it," he says. "We had a lull in the schedule."

We discuss how the harvest is going and what the numbers look like. Three key metrics determine whether the grapes are ready to pick: the Brix scale, or the measure of the sugar content; pH; and tartaric acid. The Brix should be around 24 to 25, pH around 3.2 to 3.5, and tartaric acid around 0.7. The closer a vineyard gets to that range, the better. If the numbers are off, vintners may have to add sugar or acid, compromising the quality of the fruit.

The Cabernet I'm getting has a Brix of 25, pH of 3.14, and tartaric acid of 0.75. These numbers are very good, but this year's harvest was limited due to high heat, which can inhibit ripening and give grapes a "cooked" flavor. We won't be able to get as much fruit as we wanted; Ryan can only sell us 800 pounds of Cabernet. We normally get around 1,200 pounds, enough to fill one of our 60-gallon barrels with wine leftover for topping the barrel over the year. We'll have to blend the Cabernet with Merlot or Syrah. Such is the nature of the craft.

We walk over to inspect the two bins of Cabernet that have just been picked. The grapes glisten like dark purple amethysts, smelling sweet and fragrant, attracting dive-bombing yellow jackets. The clusters are small, promising intense flavors. I pick up a handful and eat them. They have a deep, rich complex taste combining plums, currants, and blackberries, way beyond table grapes such as those going into Welch's juice. I chew them slowly, analyzing the layers of flavors and imagining the excellence of the wine to come.

Ryan brings the forklift over to dump the glistening clusters of grapes into the back of the Beast. Then he gives me the invoice, and we shake hands.

I drive slowly down the road and turn west toward I-90. Now I have to coax the creaking, squealing heap of metal back to Seattle. Once I merge onto the freeway, I pat its dashboard, encouraging it. By the time I arrive in Seattle, I've lost several pounds in sweat. Tom and the others help me unload the fruit. I tell the story of the blowout and how I barely made it into Benton City.

"You have to baby it," Tom says. He loves his truck and doesn't take criticism of it lightly.

"I'm glad I survived it," I say, vowing *never* to pick up grapes with that heap again.

Tom and I, as partners in the Les Copains wine co-op, complement each other well. Tom has a PhD in chemical engineering and a mad-scientist gleam in his eye. He loves the romance of wine and fantasizes about starting our own winery.

By contrast, I'm a French major who got bitten by the wine bug during a junior year abroad program in France. I've gone on to write about wine, obsess about wine, and make pilgrimages to the most famous wineries in Bordeaux, Burgundy, and Champagne in France; the Piedmont and Tuscany in Italy; the Mosel Valley in Germany; and the Napa and Willamette Valleys in the United States. I love the physical, sensual nature of wine making and its deep connection to the past, but I understand the business well enough to know that wine making is not an easy way to make a living.

Tom and I have our spats, but we work through them. It's all part of making wine.

The members of our Les Copains wine group assemble in Tom's driveway, ready to crush this year's crop of grapes. They shed their identities as executives, teachers, writers, and engineers, and don jeans and T-shirts to become winemakers, drawn to a process that goes back thousands of years. The activity has deep roots, stretching back to the biblical bedrock of our culture and beyond; the oldest wine-making equipment was discovered in Georgia, dating back eight thousand years. During the wedding feast at Cana, Jesus proved himself a damned fine winemaker, as well as the Son of God, by turning the water into wine and producing even tastier juice than the first round. Wine has served as a symbol and sacrament throughout the pagan and early Christian world. All of this history adds to the magic of the process.

After a ritual grape stomp, our crush gets underway. Two of our buddies scoop grapes into a big plastic bucket and carry it over to the stemmer/crusher. The twenty-year-old machine is dented, rusted, and stained with grape juice but has served us well over the last ten years. They lift the hundred-pound bucket and tip it into the crusher. Tom guides the grapes into the hopper on top of the crusher. Theresa Jurotich, a longtime *copaine*, controls the switch. She flips it on, and the machine springs to life. The grapes disappear into the gleaming stainless steel maw of the machine. Two rotating drums crush the grapes and send them past a spinning steel rod surrounded with tines that extract the stems before they fall into the plastic fermenting bin. The machine groans and squeals and complains like the Beast. Bright purple juice spills into the fermenting tote. Fruit flies circle excitedly. We taste the juice and toast the harvest.

Then the chain slips off the machine. We take off the chain, reattach it, and start it up. It falls off again. Finally, Scott Driscoll, a writer friend, finds a two-by-four to brace against the gears so they don't slip out of alignment and throw the chain. The smell of burning wood fills the air.

I try to relax but worry what will happen if the stemmer/crusher breaks down. Sometimes our operation seems to be held together with duct tape and Elmer's glue, as the contraption is perpetually on the verge of a breakdown. I started making wine fifteen years ago, and the hobby has continued to grow like a kudzu vine. My wife, Lisa, used to park her car in our garage. Now the garage houses four sixty-gallon French oak wine barrels, two thirty-gallon barrels, and dozens of glass carboys. It's a small winery crammed into the back of the garage. Lisa has been gracious so far, but the "hobby" seems to keep growing.

And it isn't cheap. Every year, we buy one new French oak wine barrel at $1,000 a pop; the three others are of varying ages. The grapes cost $3,000 per ton, and then the various other expenses, such as sulfites, bungs, yeast, bottles, corks, and truck tires, add up. The payoff, beyond the process itself, is the wonderful juice. We produce eight hundred bottles of rich, full-bodied reds: Cabernet, Merlot, and Syrah. Though our wines can be inconsistent, we often make stellar wine.

People taste it and ask, "When are you going to open your own winery?"

"I don't want to open a winery," I reply, keeping in mind the joke that opening a winery is a great way to make a small fortune out of a large fortune.

I have no large fortune to spend. I have a career as a writer and teacher. I'm in my fifties with a wife and three kids. I don't need another job. I don't need another obsession. But sometimes you don't choose your obsessions. They choose you.

As we near the end of crush, I hoist the last bucket into the hopper. Tom guides the grapes into the machine. Pants and shirt stained purple from loading grapes, I dip a cup into the must, anticipating its taste and texture after the alchemy of yeast has transformed it into the miracle of wine.

"This is fantastic juice," Tom says.

"It's amazing," I say, "even at this stage." The must has a balance between sweetness and acidity, and layer upon layer of flavors: blackberries, plums, orange peels, camphor, herbs, and exotic spices. I hold a universe of tastes in a small plastic cup.

My resolve is beginning to crumble. Is this my midlife crisis? I shake my head; this is insane. But then I taste the wine again and imagine what it could be like if I knew what I was doing. I've tasted some of the best wines in the world: Gaja, Château Margaux, Château Latour, Château Lafite Rothschild, Château d'Yquem, Cheval Blanc, Domaine de la Romanée-Conti, Stag's Leap Wine Cellars, Quilceda Creek. I know their richness and refinements, the complexity of their flavors, the uncanny sense of place and character and culture they evoke, the mystical Dionysian spirit they conjure. Could I actually make wine like that? The fruit is there, but the craft, alas, is not.

I help Tom hose out the stemmer/crusher. It's been a long day, but I can't stop sipping the must. Something about it represents a challenge, an irresistible possibility. Barrel fever, some call it.

I put the cup down. If I want to take this further, I need to know more. I have a grasp of the fundamentals of wine making, not the refinements. I've read books about it, but I need to take the next step. I need mentors, guides in the art and science of wine making. But who could I ask?

2

Ciel du Cheval

A Wine Lover's Idea of Heaven

Red Mountain rises above the low, brown sagebrush country outside of Benton City, Washington. Located just northwest of the Tri-Cities, Red Mountain is one of Washington's smallest and most prestigious American Viticulture Areas (AVAs). One of a series of rounded basaltic ridges marching south along the Yakima River, it looks more like a hill than a mountain.

The landscape is bare, dry, austere. It's nothing like the carefully tended gardens of Burgundy or the regal architectural jewel boxes of Bordeaux's first-growth Château Latour.

Approaching Red Mountain in my Eurovan (thoroughly customized by my kids but reliable nonetheless), I don't have to worry as I did about the Beast breaking down, so I have time to take in the view of green vine rows undulating toward the top of the rocky summit. In addition to picking up our Merlot grapes, I've scheduled a tour of Ciel du Cheval Vineyard with owner Jim Holmes. I'm hoping to learn how this site grows such fantastic fruit and use that knowledge to improve my wine making. They say that wine is made in the vineyard. If I want to make great wine, I have to learn what that means.

Like many winemakers in Washington, I purchase grapes rather than growing them myself. Once upon a time, I dreamed of having my own vineyard, which led me to plant six Pinot Noir vines on the south side of my house. The cool, rainy Seattle weather yielded a quality harvest only one year out of five. One year, the weather stayed warm and sunny throughout the summer. The grapes hung fat and juicy from the vines. I planned to pick them on a certain day, but the evening before, a rainstorm moved in, inundating the grapes. The grapes absorbed the water, splitting the skins and causing rot. The crop was ruined.

After that reality check, I began purchasing fruit from the dry eastern side of the state. I joined a group of amateur winemakers buying grapes from Sagemoor Vineyards, an excellent site. I found the resulting wine good but with a green pepper character not to my taste, so I kept looking for another vineyard. Eventually I found my way to Ciel du Cheval, where the elegance and balance of the fruit seemed perfect.

How does Ciel achieve this? How do the climate, soil, and personality of the grower influence the grapes coming out of the vineyard? These are some of the questions I need to answer to make great, site-specific wine.

The road to Red Mountain follows the meanders of the Yakima River, a slow, sluggish stream in the fall. I turn left and head up the road toward Ciel du Cheval Vineyards. Rows of grape vines line the road. Carefully tended clusters of Cabernet, Merlot, and Syrah grapes dangle from beneath the vines. Today, black clouds loom to the north, but it's dry—for now—on Red Mountain.

I turn right at a gravel road next to a line of poplar trees and enter Ciel du Cheval Vineyard. I'm slated to pick up six hundred pounds of Merlot. I'm hoping the rain hasn't damaged or limited the crop. Grape growing is an unpredictable business; too much heat, cold, or rain can yield a small crop. Ciel then parcels out what's left, sometimes partially fulfilling the buyers' orders. I'm hoping that doesn't happen today.

The air is thick with the sweet smell of ripe grapes. I park near the outbuilding, a boxy wooden structure amid the vines. Jim Holmes and vineyard manager Ryan Johnson are busy discussing how to use a smart phone to track clouds approaching the vineyard.

"We're getting about a tenth of the annual rainfall today," Ryan says, looking up from the phone.

"See, the old guy is teaching the young guy how to do it," Jim says, showing me the screen, which indicates the position of clouds over the vineyard, giving him very precise data on when the rains might arrive and how to schedule the picking. The vineyard uses an extensive system of sensors to monitor the weather and water in the soil.

A pioneer of the Washington wine industry, Jim is a sturdy man with a red face and thick, expressive lips who wears jeans, a jean jacket, and a University of California (UC) at Berkeley Bears hat. Every harvest is different, and Jim always has the pulse of it. He sends out detailed emails prior to the harvest about the ripeness of the grapes, including Brix, tartaric acid, and pH levels—the three essential chemical measurements in grape maturity that indicate when to pick.

Ciel's professionalism and high-quality fruit attract the leading vintners in the state. Getting on the list to purchase grapes here is very difficult. Like so many other things in the wine industry, grape buying is extremely competitive.

"We take care of all of them," Jim says of his customers. "We've got forty-one blocks, each managed differently."

Now in his late seventies, Jim still enjoys the challenge. A systems engineer who grew up in the Bay Area north of San Francisco, he developed an early taste for wine. When he moved to the Tri-Cities area in 1958 to work for the Hanford Nuclear Reservation, he wanted to drink some decent local wine. The wine available was Tokay, a high-alcohol dessert wine. He wanted something more. So he and his partner, John Williams, bought eighty acres of land on Red Mountain in 1972. Shortly thereafter, Washington State University professor Walter Clore released his groundbreaking report on Washington wine. It argued that *Vitis vinifera* grapes, which produce high-quality wine, could grow in Washington.

"Did the Clore report inspire you to plant grapes?" I ask. I've heard this but wanted to hear it from Jim directly.

"It was a great report," Jim says. "He'd done twenty years of research. I tasted some of the wine, university wine. Mostly it was atrocious, but some of it was good. It turned out to be the start of the whole thing."

Clore's data convinced Jim that growing grapes was worth doing. "It was a high-level risk," he says, gesturing toward the vineyard. "Anyone who planted grapes on Red Mountain back then could qualify as crazy."

Despite the risk, Jim began planting vines. In 1975, his first year of production, they yielded sixty tons. He planned to sell them to home winemakers, but Preston Cellars was looking for fruit. He sold it all to them. He's had no problem selling his fruit since.

When I ask to see the vineyard, he offers to give me a tour in his white Ford Expedition. "Hop in," he says. He drives slowly and deliberately down a road between the vines, wheels crunching over the gravel. The vineyard includes 120 acres with an additional forty acres jointly owned by Jim, Quilceda Creek Vintners, and DeLille Cellars. All the rows are carefully pruned and numbered so that Jim can track the progress of the harvest.

"The flavors are great this year," he says, as he turns down another gravel road, the truck kicking up dust. "That's because it's been cooler. We haven't had the heat we normally have. That's allowed the flavors to develop. If it's too hot, it's dangerous to the grapes. We haven't had a year like this since 1999, and that was a very good year."

We pass row upon row of vines and gentle undulations of the landscape, reminiscent of vineyards in Bordeaux, which are far from flat but not as steep as the vineyards in Burgundy or those in Germany's Mosel Valley. It takes several minutes to reach the far end of the vineyard, where we park and get out of the SUV. "See the four canes coming out of the ground?" he says, pointing to the vines. "That's the trellising system we use here. Seems to get good ripening for the grapes. It's an old Russian system. It's vertical, so everything is exposed to sunlight."

I frantically take notes, trying to get all this down. My brain is getting overwhelmed with details, but I'll sort them out later. I just want to learn as much as I can right now. I shove the notebook back in the pocket of my shorts and follow after him.

Walking the vine rows, he kicks the dirt, explaining that the sandy, porous soil provides ideal drainage for the vines. This is very different from the claylike soil in my backyard, which trapped rainwater, ensuring the vines would absorb it and split the grape skins.

"The soil here has a high pH content, which makes it difficult for the vines to extract nutrients from it, forcing the vines to struggle and keeping the leafy canopy down," he says. This yields a richer and more complex wine, part of the reason I find these grapes so appealing.

As we keep walking, I notice that Jim allows his vineyards to grow in a leafy, unruly style that is more like an English garden than the crisper, cleaner lines of his neighbor, Klipsun, another storied vineyard. Every grape grower seems to have his or her own style, another element of terroir. It's not just the climate and soil but also the person who cultivates it.

"They make a more masculine style wine," he said. "Theirs is more acid, strong. Ours tends to be more elegant."

Even though the vineyards are separated by no more than a few hundred yards, he explains their soils are quite different. Some thirteen thousand to fifteen thousand years ago, the great Missoula or Spokane floods raced across the state, depositing a wide range of soils that resulted in very different characters in the vineyards. Like a university professor, Jim gives a discourse on the various kinds of silt here that drain well but also hold enough water to make it easy to irrigate.

Walking out among the vines, he leans down and picks up a tennis ball–size rock to illustrate the importance of the area's geology. "They washed down with the Spokane floods," he says of the rough and rounded stone, many of which underlie the vineyard. I grab a rock and examine it. You can't get any closer to terroir than this, studying the soil and the stones that make up the vineyard—all key to understanding how to make great wine.

We keep walking the rows, kicking up the dust that coats the lower leaves of the vines. Jim checks the size and color of the grapes and leaves to see if they're getting enough water. The temperatures are cooler than expected, but it's still hot and dry. I wipe the sweat from my brow as I follow him down the rows.

In addition to soil and trellising, climate plays a key role in grape production. Ciel du Cheval, like the rest of Red Mountain, is a desert—arid, hot, flooded with sunlight—which is ideal for growing wine grapes. The soil and climate make it an outstanding site, unlike the backyard of my Seattle home.

"It is a warm site," says Jim, "typically running three thousand degree days."

"What's a degree-day?" I ask. I thought I knew a lot about wine, but many aspects of running a vineyard are new to me. I keep taking notes, trying to get all of this information down.

He explains that his degree-day measurement represents the cumulative amount of heat the grapevine experiences during the growing season, which runs from April through late October. Degree-days determine whether a site can ripen a specific type of grape. Generally, the higher the number the warmer the climate and the better for ripening. At Ciel du Cheval, the hot and sunny climate allows him to ripen many different varietals.

However, too much heat can damage the fruit. If the air temperature rises above ninety-seven degrees Fahrenheit, the grapevine may stop metabolizing and become damaged. This danger has been negligible at Ciel du Cheval, fortunately. "Generally, peak temperatures seldom get above one hundred degrees Fahrenheit," he says.

He explains that temperature variation also plays a part in producing outstanding fruit. The difference between the high daytime temperatures and the cool nights is as much as forty-five degrees. The cool nights develop flavors and character in the grapes by slowing the ripening process and by maintaining the crisp acidity—one of the key ingredients in making a balanced wine.

Fully ripe grapes with the right level of acidity exhibit balance, elegance, refinement—all the qualities I crave. I have no desire to make a "fruit bomb," a massive wine high in alcohol that can earn high scores from critics simply because it's big, showy, and bombastic but does not pair with food or reveal the nuances of the vineyard. Such wines have become quite popular in recent years, but I prefer the balance and elegance of Ciel fruit. I don't want to make a heavy metal wine. I want to make a Beethoven symphony of a wine. It's exciting to see how I might be able to do it with grapes from this vineyard.

The right climate and soil are critical in a vineyard, but they only go so far in explaining Ciel's success. Jim modestly omits his own role, but the human factor is central to any discussion of terroir. Ciel has thrived because of Jim's drive to grow the best fruit, constantly tinkering with the varietals, finding the best clones, experimenting with the irrigation system, always willing to adapt.

Ciel may be mature by the standards of Washington, but it is an upstart compared to some of the world's great vineyards such as those of France's Domaine du Vieux Télégraphe, which I've been lucky enough to have visited several times. Those vineyards are hundreds of years old, with detailed records of how they have performed over a range of conditions. In more than thirty years of grape cultivation, Jim has learned much about his vineyard, but he admits there is much more to know. For now he simply concentrates on what works. He once planted Chardonnay, Riesling, and Gewurztraminer, but he replaced them with Cabernet, Merlot, Syrah, and newer varietals such as Brunello whose potential interests him.

We keep walking, checking on the fruit.

"We're concentrating on what we do well," says Jim. "Things that don't do well, we pull out!" Red Bordeaux varietals such as Cabernet and Merlot have become standouts at Ciel, earning exceptional scores from wine critics. Rhône grapes, both red and white, have also done well, with some speculating Syrah may eventually be the varietal best suited to the site. Meanwhile, Jim continues to experiment, using a variety of trellis styles—from the standard vertical shoot position to a less common fan trellis—seeking the ideal system for his vineyard.

"Making great wine is like painting a great picture: You have to have some idea of what you want to paint before you put paint to canvas," he says, fingering a cluster of grapes. "We've got a great site, we've got great paint! If you want to make a great bottle of wine, you've got to think about what you're doing in the field. You've got to know why you did this, and why you did that, and why one worked and the other didn't work. You've got to keep working at those things because the desire is to put great art on the canvas."

This is exactly what I need to know. I have the desire to make great wine. I'm getting great fruit. But how can I make a beautiful canvas of it? I'm beyond the paint-by-numbers approach, but I'm nowhere near a Cézanne or Picasso version of a wine. There's a big difference between a passionate amateur and a seasoned professional. I'll really have to up my game if I want to get the most out of the Ciel fruit. Today's tour represents a step in that direction, in my getting to know in detail how Jim grows the grapes so that I will know better how to work with them.

"It's still scary," he says of grape growing as we walk back to the SUV. "The first frost can damage the plants. Bad weather, especially during the harvest, can make or break a year's crop." In January 1996, temperatures plummeted as low as minus eighteen degrees in eastern Washington, causing widespread damage to the vines, more than halving the crop, and nearly putting some wineries out of business. In other years, rain or excessive heat has damaged the crop. Today he keeps a close eye on the dark clouds to the north, remaining on high alert until the last grape is plucked.

"When the last truck rolls down the road with the last load of grapes, the referee can blow the whistle, and the game is over," he says. "You're relieved. You look at all the bullets you dodged during the year. Nothing is prettier than a picked vineyard."

He drives me back to the loading area. I spot my six hundred pounds of Merlot sitting in picking bins next to my van. I thank him for the tour and examine the grapes. As always, they're clean, ripe, fat clusters with few leaves among them. With his and others' help, I hope to make wine worthy of them. I pop a grape into my mouth, savoring the deep, rich, complex flavor, envisioning the wine it might become. Now if I can just develop the skill to make that happen.

PART 2

Washington

3

The Founding Fathers of the Washington Wine Industry

Terroir is often considered simply dirt and climate. But after visiting world-famous wine regions such as France, Germany, and Italy, I saw clearly that the people who helped found and develop them were an integral part of the terroir: the Perrin and Brunier clans in Châteauneuf-du-Pape, Ernest Loosen in the Mosel Valley, and the Biondi-Santis in Tuscany's Brunello region.

For this reason, I wanted to deepen my enological education by seeking out the founders of the Washington wine industry. I hoped to learn more about why they got into the industry and how they learned their craft in hopes I could make better juice myself.

Washington is the second-largest wine-producing state in the United States and is quickly earning recognition as one of the premier grape-growing areas of the globe. It boasts more than a thousand wineries, four hundred grape growers, and sixty thousand acres of wine grapes. It cultivates eighty varieties of grapes in nineteen different American Viticultural Areas. It has a yearly in-state economic impact of more than $8 billion and produces over seventeen million cases of wine.

This success seems inevitable now, but it didn't in the early years. It happened because its leaders had a frontier spirit, a willingness to cooperate, a penchant for experimentation, and a fanatical dedication to quality: Walter Clore of Washington State University (WSU), Jim Holmes of Ciel du Cheval, John Williams and family of Kiona Vineyards and Winery, David Lake of Columbia Winery, Norm McKibben of Pepper Bridge Winery, Gary Figgins and Nancy Figgins of Leonetti Cellar, and Rick Small of Woodward Canyon Winery, as well as the indispensable company Chateau Ste. Michelle, which played a crucial role in the growth and development of the region. Why did they get into the industry? How did they overcome the many obstacles of the early years? What can I learn from them in my own wine making?

The Johnny Grape Seed of Washington

One of the founding fathers of the state's wine industry didn't drink. Walter Clore, a horticulturalist who spent his entire career at the WSU Irrigated Agriculture Research and Extension Center near Prosser, Washington, had no taste for alcohol but loved propagating fruits, berries, asparagus, and grapes. He is credited for discovering that European *vinifera* could grow in the vast, semidesert of eastern Washington, a region where water is scarce, summers are hot, and occasional arctic blasts devastate the vines.

"His impact is huge," says Mike Sauer, who with Clore's guidance started growing wine grapes in the late 1960s at his now-acclaimed Red Willow Vineyard in the Yakima Valley. "As one of nature's gentlemen, he's always shown kindness to whoever is seeking information about growing grapes. That struck me, as a young man who just wanted to plant a few vines in the beginning. He took me seriously, perhaps more seriously than I took myself. Walter has inspired at least a couple of generations of grape growers."

Born in Tecumseh, Oklahoma, in 1911, his family moved to Tulsa, where visits to his grandparents' farm instilled an early interest in agriculture. After graduating from Oklahoma State University with a degree in horticulture, he did odd jobs before winning a horticulture fellowship in 1934 at Washington State College in Pullman that paid $500 a year.

In 1937 he became an assistant horticulturist at the Irrigated Agriculture Research and Extension Center. This dream job allowed him to research possible crops for the vast Columbia Basin. He cultivated apples, grapes, raspberries, and blackberries to learn what crops flourished in the dry sagebrush country.

"When I first arrived at the Experiment Station in Prosser, I felt like a pioneer," Walt says. "With one million acres of the Columbia Basin to be irrigated, and the prospects of seventy-two thousand acres on the Roza [Irrigation District], I felt that I had to learn about the adaptability of many horticultural crops."

Most important, he helped plant twenty American type grapes, mostly *Vitis labrusca* hybrids, as well as seven *Vitis vinifera* grape varieties. It was the beginning of his long association with grapes and their culture.

The station's "mother block" of trial vines expanded over the years, including American hybrids, French hybrids, and *vinifera* grapes. Most of the stock came from California, but William B. Bridgman's Sunnyside vineyard contributed some *vinifera* grapes. Bridgman, an attorney, started Upland Winery in 1934 and served as a mentor to Clore.

Watching over the block carefully, Clore cataloged the effects of weather, temperature, fertilizers, irrigation, ground cover, trellising, and pruning methods. He traveled around the region, studying the soils, microclimates, and locations that yielded the best fruit.

These studies revealed premium *vinifera* grapes could thrive in Washington, provided growers selected the correct sites. His rule of thumb was, "If you can't grow Riesling and get it to survive, then you shouldn't be planting anything."

Winemakers and growers consulted with Clore and fellow researchers Charles Nagel and George Carter to determine where and what to plant. In addition to working with the wine industry, Clore provided crucial testimony to the state legislature about the need to update the state's wine laws.

In 1969 the state legislature held hearings about whether to overturn the state's protectionist wine laws that favored Concord grape production. Clore testified that Washington could no longer compete with California in producing Concord table grapes for jelly and juice but could compete in the growing fine wine market. He explained that the state shared the same latitude as the famous growing regions of Europe and had a longer growing season, more hours of intense sunlight, and fewer insects and diseases than California did.

Clore's testimony proved decisive in convincing the legislature to open Washington's wine market, paving the way for its wine industry to focus on premium grapes rather than grapes for juice or jelly. While most academic research gathers dust in a library, Clore's findings helped jump-start the state's fine wine industry. His testimony and research piqued the interest of many of the state's winemakers and grape growers, including Ciel du Cheval's Jim Holmes.

Jim Holmes of Ciel du Cheval on Washington's Red Mountain

"Anyone who planted grapes on Red Mountain could qualify as crazy," said Jim Holmes of his and John Williams's decision to plant vines in 1975 on the deserted, windswept side of Red Mountain located outside the Tri-Cities.

While working as a research manager for Westinghouse at the Hanford Nuclear Reservation, Holmes met fellow engineer John Williams. Both had invested in the stock market and lost money; now they planned to invest in real estate and make a killing. Williams's father-in-law owned property on Red Mountain. They purchased eighty acres of it for $200 an acre, intending to turn it into a shopping mall.

The area looked like a wasteland, dotted with tufts of sagebrush, scoured by wind-blown silt, and inhabited by bony jackrabbits. It had no road access. No electrical supply. No water. It was like walking on the moon.

But Holmes's upbringing in the Bay Area and travels to Europe had piqued his interest in wine. In 1969 WSU's Walter Clore presented a report to the Washington State legislature that argued growing quality wine grapes in the state was possible. His subsequent research at the university's Irrigated Agriculture Research and Extension Center in Prosser demonstrated how to successfully grow *vinifera* grapes, the basis for the world's finest wines, even in such an apparently unpromising location as Red Mountain.

After reading Clore's report, Holmes and Williams started putting money into developing their property. They brought in electricity, dug a well, and planted ten acres of grapes, including Riesling, Chardonnay, and Cabernet Sauvignon.

With the industry in its infancy, they had no idea of what to do with the grapes. Eventually they contracted with nearby Preston Cellars, where winemaker Rob Griffin had made a name for himself. The first harvest of grapes went to Preston in 1978; the resulting Cabernet proved stunning.

The land rush was on. The original eighty acres became Kiona Vineyards. Some friends purchased the adjoining eighty acres on Red Mountain that later became Ciel du Cheval vineyard. In 1980 Holmes and Williams started producing wine under the Kiona label. Williams moved to the Kiona land in 1983.

In 1991 the Kiona partners purchased the promising Ciel du Cheval vineyard. Holmes and Williams enjoyed working together, but in 1994 they decided to divide the business for estate reasons. Williams got the Kiona

Vineyard, while Holmes took over Ciel du Cheval, which was only partially planted. This allowed him to experiment with new varieties.

Holmes retired from his engineering job that same year to focus on Ciel du Cheval. He brought his scientific background to grape growing, applying new training and growing techniques to make the best of the locale.

"I'm still a scientist, so I've been working on data that we've collected here just by accident. Again, because I was an engineer, this place is highly automated. We have automatic watering systems; the soil moisture has been recorded in a database for tens of years. We have an immense amount of data, which together with WSU and Oregon State University, reveal some previously unknown relationships between soil and vine."

His focus on analyzing data allows Ciel to continue to improve. He is modest and self-effacing, but the awards keep stacking up, with numerous hundred-point Robert Parker scores. Many of Washington's leading winemakers covet his grapes, and he sells to about twenty-five of them, including me. His son Richard runs the business side, and another son, Kade, runs the vineyard operations. In 2007 Jim Holmes and his wife Pat were inducted into the Washington Wine Hall of Fame.

In the years I've purchased fruit from Ciel, I've learned a ton from Jim. Every time I pick up fruit, he gives me a tour, explaining how the harvest went. No harvest is the same. Sometimes excessive heat limits the crop. Other times cold temperatures threaten the vines. Richard emails me up-to-date numbers on Brix, pH, and total acidity levels, as well as information about who is picking. Since I know most of the other winemakers, this guides me in deciding when to pick. As noted previously, pick too early, the wine can be green and vegetal. Pick too late, it's an alcoholic fruit bomb. Like Goldilock's choice, the Ciel team helps me get it just right.

All in the Family: Kiona Vineyards and Winery

No family can claim responsibility for the creation of Red Mountain quite like the Williams clan, owners of Kiona Vineyards and Winery. Their creation story began when John Williams purchased property with Jim Holmes on a desolate patch of dirt outside the Tri-Cities and planted a vineyard in 1975 that would later become part of the Red Mountain AVA.

"My grandfather decided to pour his life savings into planting a twelve-acre vineyard," said JJ Williams, general manager of the property. "He had to bring in power, lay down a gravel road, and dig a well. People thought he was nuts."

Kiona first planted Cabernet, Riesling, and Chardonnay. What began as John Williams and Jim Holmes's pet project soon became a family affair. John's son, Scott Williams, recalls the early days of working the vineyard.

"I was in high school and I recruited a bunch of my classmates to help us plant the vineyard. There must have been a recent gnat hatch, because on one really hot Saturday afternoon there was a giant swarm. They would fly into your nose and mouth, and it was hard to see. Everybody wanted to go home, but I told them that we had to finish. They started to sing 'Chain Gang' by Sam Cooke. I felt bad after that, so I sent them all home!"

The others left, but Scott and the rest of the Williams clan kept working, gradually expanding and improving the site. It had great potential, with sandy loam providing excellent drainage for the vines. Scott recalls driving though the vineyard in 1977.

"We would drive to the vineyard site in a 1969 Plymouth Fury station wagon. It was yellow with black vinyl interior. The road was so bad at the start. It was really just a path through the desert. There were these soft dunes made of ultra-fine dust that you'd have to build up speed to bust through or else you'd get stuck. So we'd gun it, and each time we'd hit another dust dune, it would send up a giant plume. It got everywhere; it came in through the vents in the car. It was like throwing flour in someone's face, except dirt."

All that fine dust made driving difficult but proved ideal ground for premium grapes, both white and red varietals. But with the industry still in its infancy, the family didn't know what to do with the production.

"Way back when, it was a fool's journey," said Scott. "There was no market, but we had to keep going forward. We're still kicking, which all by itself is quite an achievement."

They had planned to sell the first Red Mountain Kiona Cabernet Sauvignon to garagistes, but winemaker Rob Griffin of Preston Cellars, now of Barnard Griffin Winery, purchased all of it.

"The conventional wisdom in the late 1970s was that Washington was a first-class white wine region with limited prospects for reds. My opinion on this point was permanently changed in 1978 with the opportunity to crush the first crop from Kiona Vineyards on Red Mountain. The depth of color and fruit intensity were definitely a revelation as to the potential for Washington Merlots and Cabernets. The fruit yielded wines of tremendous depth and intensity, real diamonds in the rough and a foreshadow of great things to come."

Kiona expanded from the original 12 acres to today's 272-acre property. Scott grows the grapes and serves as the fix-it guy for the property. Early on, he involved his family in farming and running the business. That paid off when his son JJ became sales manager in 2009 and son Tyler became the winemaker in 2019.

"I'm glad I have JJ," Scott said. "Growing up he expressed interest in the business. I told him when he went off to school, 'We have a yahoo farmer and an engineer in the family. We need a businessman.' He started out doing the sales and marketing. Now he runs the business."

Growing up on Red Mountain, JJ gained a unique insight into the business of growing grapes and making wine. At age fourteen, he began working for his dad, Scott Williams, and eventually served as the youngest member of the Red Mountain AVA.

"You could take ten steps out the front door, and you'd be in a vineyard," he said. "It's a heck of a way to grow up. For others being in a vineyard is so romantic, but for us, it was digging a lot of ditches."

Tyler also expressed an interest in Kiona. After receiving a degree in biology and a master's in enology, he traveled the world, apprenticing at wineries in Bordeaux, Sicily, South Africa, Chile, New Zealand, Australia, and several top-tier Washington houses.

"Before I gave him [Tyler] the keys to the car, I had him go back to school for his master's," Scott said. "He had to earn it. Handing the keys to the car is just that. You've got your license. Let her rip. I don't tell him what to do."

The family meshes well together, with each of the members contributing to the effort, which comes to fruition in bottles like their 2016 Estate Red Mountain Reserve. "The idea behind the wine is to designate geographically, not by grape type," said JJ. "It's a blended wine from this place. It's the best wine we can make in a given vintage."

Such a wine grows out of the time, sweat, and toil the Williamses have put into making the Red Mountain AVA a world-class vineyard site.

"We're taking the long view on the business," Scott said. "We own it outright. We want to be around in a hundred years."

Judging from the quality of the fruit, they're on track to do this. I remember well the thousand pounds of Cabernet I purchased from Scott. The resulting wine was sleek, stylish, and delicious—one of the best wines Les Copains has ever made.

To Oak or Not to Oak? Hightower Cellars

To oak or not to oak? That is the question. I'm sampling the wares of Hightower Cellars, a lovely hillside vineyard and tasting room on Red Mountain in the southern end of the Yakima Valley. Tim Hightower, a co-owner, pours a sample of his offerings into a crystal glass. I swirl, sniff, taste, and—regretfully—spit.

I savor the crisp, vibrant flavors of the wines that resemble black cherries, blackberries, and dark plums, with no smoky barrel flavor to obscure the subtle aromas and tastes. "There isn't much oak in these wines," I say, putting my glass down to taste the next sample.

Hightower, a friendly man wearing a wool sweater and dungarees, just smiles. "We like oak as an ingredient," he says, "but don't want it to overwhelm the wine."

For years, winemakers have aged wine in new oak barrels, seeking wines with smoky wood flavors. While these wines can taste great, such treatment doesn't work for every wine. New oak can obscure the individual characteristics of the vineyard. Recently, some winemakers have cut back on new oak, using older, neutral barrels that let the subtleties and nuances of the vineyard emerge. Hightower is part of that vanguard, adding yet another stylistic twist to the rapidly changing world of Washington wine making.

Founded in 1997, Hightower is one of the newer wineries in the Red Mountain AVA. Located northwest of the Tri-Cities, the Red Mountain AVA is part of the Yakima Valley AVA, which in turn is part of the larger Columbia Valley AVA. Located between Benton City and Richland, the Red Mountain AVA is tiny (only 4,040 acres) by comparison to other state AVAs, but its reputation looms large.

Many of Washington's boutique wines come from Cabernet Sauvignon grapes grown here, including the 2002, 2003, 2005, and 2007 Quilceda Creek Vintners' Cabernet Sauvignon, which earned rare hundred-point ratings from Robert Parker of *The Wine Advocate*. Only fifteen other wines in the United States have received this designation, all made from California grapes. The Quilceda Creek wines were blends from mostly Red Mountain vineyards, hence its worldwide reputation.

Red Mountain is a warm, dry site with well-drained soil perfect for growing red wine grapes such as Cabernet Sauvignon, Merlot, Cabernet Franc, Syrah, and Petit Verdot. Recently, growers have added Sangiovese and Malbec to the mix. The AVA excels at Bordeaux varietals—Cabernet, Merlot, Cabernet Franc—yielding rich, powerful, tannic wines, and many of them marry well with new oak. Thus, it was a surprise to visit Hightower and see a departure from this trend.

Hightower, also an excellent amateur chef, credits his culinary background with making wines that pair well with meals. "We like to drink wine with food," he says. "We use less oak and try for wines that are more refined."

Other winemakers might prefer a tannic, oaky Cabernet to pair with steak, for example, but the important thing is that there are now options. The stylistic diversity of Washington wines continues to increase.

The Tri-Cities

To round out my knowledge of Red Mountain, I visit some of the surrounding wineries in the nearby Tri-Cities. Like other areas of the state, the Tri-Cities region has greatly enhanced the wine-touring experience. Barnard Griffin Winery is the latest house to expand its tasting room and open an on-premises restaurant. Located on the west bank of the Columbia River off I-82, the Kitchen at Barnard Griffin's joins neighbors Bookwalter and Tagaris wineries in pairing fine cuisine with their wines.

"We realized we needed to expand the visitor experience," says proprietor and winemaker Rob Griffin as he shows me around the new restaurant decorated with brightly colored glass art in the form of giant yeast cells that his wife, Deborah Barnard, designed and handcrafted. "We ask people if they can guess what the shapes represent. Only a couple of winemakers have guessed right."

A wine geek from way back, Griffin graduated from the enology program at UC Davis before migrating north to take a job at Preston Cellars in 1977. He's witnessed firsthand the exponential growth of the state's wine industry.

"When I started there were fewer than ten wineries in the state," he says. "It was always missionary work to sell the wine. People would say, 'I don't know why they grow grapes up there.' In 1978 Ray's Boathouse chose us as their house Chardonnay. That was the start of it."

Things took off from there, both for the region and for Griffin, as he quickly established a reputation for making clean, bright wines brimming with beautiful Washington fruit. The winery has won many awards, including a 2013 San Francisco Chronicle Wine Competition's Best of Class for the 2011 Chardonnay.

Next door, J. Bookwalter Winery has also expanded the wine-touring experience with its tours, tastings, and stylish bistro Fiction—all of which draw increasing numbers of visitors. Bookwalter Winey, which specializes in red blends and Riesling, houses an ideal spot to pair food and wine. For many years, the region has lacked places to try wines with local cuisine. That is changing for the better, with Bookwalter at the center of that synergy.

Bookwalter sources grapes from renowned vineyards such as Conner Lee in Walla Walla and Elephant Mountain near Wapato. This gives the winery flexibility and control in its wine making and blending.

"Nothing but the best fruit goes into our wines," says Travis Maple, assistant winemaker. "We use a single-berry sorting system. Only the best of the best of the best grapes goes into the wine."

A single-berry system requires someone to inspect every grape and discard any imperfect fruit, bugs, sticks, stems, or leaves. The process is very labor intensive, but it yields pure, immaculate wines. In addition to using this system, owner John Bookwalter has tapped celebrated French winemaking consultant Claude Gros for expertise. The winery has begun keeping the wine on the lees, or sediment, until very late in the aging process. Bookwalter believes this allows the wines to open up and display fresher fruit flavors and a richer mouth feel. Just as at Hightower, the judicious use of new oak barrels makes the wines light on their feel and perfect for pairing with the dishes such as crab cakes and Wagyu beef tenderloin served at the winery's Fiction.

"The analogy I use is that oak is like the frame of a picture," says John Bookwalter. "It can accentuate the picture if it is chosen correctly, but it shouldn't get in the way of the picture, or the fruit in our case."

Literary types like me find the bookish references irresistible. Bookwalter calls his 2009 red blend Subplot No. 26; 2009 single-vineyard Conner Lee, Conflict; and 2009 blend Protagonist. They have yet to add more literary names such as denouement, but another is likely coming. Adds Bookwalter, "I have thought about using denouement, but I am not sure anyone can pronounce it, including me."

He credits the success of his wines in large part to the quality of the vineyard sites in the region. To get a better idea of what shaped them, I drive to the nearby Twin Sisters Trail, a short, scenic romp that serves as an easy introduction to the landscape formed by the Missoula floods, which poured through nearby Wallula Gap and deposited a fine layer of loess, or silt, over sites in eastern Washington. This loess, a well-drained soil ideal for growing grape vines, forms the foundation of some of the region's best vineyards, including Ciel du Cheval, Kiona, and others.

The trail heads up a draw, the air pungent with sagebrush. Then it circles around the back side of the two pillars of basalt, six hundred feet high, that are volcanic remnants after a great flood near the end of the last Ice Age, or thirteen thousand to fifteen thousand years ago. Scrambling up a chute, I emerge at a gap between the two pillars and take in a view of grassy hills and a calm Columbia River so tranquil it's hard to imagine that floodwaters once raged past here. It's a great macro view of the forces that formed the state's vineyards and how they made possible the wonderful wines coming out of the state.

David Lake and Columbia Winery

Lake watched as the forklift brought another bin of Red Willow Syrah grapes over to the crusher. The bin wobbled in the air, turned over, and tumbled the perfectly formed clusters into the plastic bins where a worm drive crushed them, stemmed them, and sent the juice, skins, and seeds into the fermentation tanks. His minions kept track of the sulfites to prevent spoilage, measured the sugar level, and supervised the early stages of fermentation.

"Twenty-three Brix," called out the assistant, indicating the sugar content of the grapes. The Syrah grapes, a variety that Lake helped introduce to Washington, will make part of his signature wine.

The crush has begun. The fever begins to build. "Every year there are new twists," Lake said. "This is the time we live for."

These grapes have been watched over, worried over, sweated over in the hot desert climate of the northern Yakima Valley. Now the result of that work will come to fruition.

Lake is guardedly optimistic. "It's a good-size crop this year," he says. "Once fermentation is in progress, we'll have a better idea. Sometimes we don't know until a year later. I've learned to be cautious about predictions."

Wine making is always an adventure, especially in Washington, where the harsh desert climate can kill off the vines. The flip side is that this harsh climate makes grapes suffer, producing a wine with impressive balance and character.

"Washington as a region has great advantages," he said. "We have remarkably good weather for ripening grapes. It's a tremendous time to be making wine in Washington."

Traveling with Lake as he toured Red Willow Vineyard, I learned that restraint and balance were watchwords for him. Pairing wine with food proved the ultimate test. This strongly influenced my wine making and gave me an early taste for Syrah, one of the signature wines of the Columbia Valley.

Described as the "Dean of Washington Winemakers" by both *Wine Spectator* and *Decanter Magazine*, Lake began his career in the wine industry in 1967 with the celebrated British wine shipper Saccone & Speed. In 1975 he became a Master of Wine, passing the combination of tastings and written tests that is widely recognized as the world's most rigorous professional examination in the art and science of wine.

David then traveled to California to further his knowledge of viticulture and enology with a year of study at UC Davis. In 1978 he moved to Oregon to work with David Lett at Eyrie Vineyards. He also worked with Amity and Bethel Heights Vineyards, acquiring invaluable practical experience.

Joining Columbia Winery in 1979, David took over the wine-making duties from Dr. Lloyd Woodburne. In his own words: "I came to Washington to explore the distinctive fruit qualities and remarkable natural balance I had noted from this state."

Among his many accomplishments in the wine industry, David is most renowned for his experimentation with new varietals and for his wine innovations. He was the first winemaker in Washington State to release a series of vineyard-designated wines and the first to produce Syrah, Cabernet Franc, and Pinot Gris in the state. David also produced the first Merlot in Washington to be blended with Cabernet Franc, thus earning the wine the distinctive name of Milestone Merlot.

In his years at Columbia Winery, he presided over the transformation of the entire industry. The puckish, low-key Lake avoided the spotlight but embodied the adventurous spirit of Washington's wine industry. Every year he turned out an amazing range of fine bottles: the inexpensive Cellarmaster's Riesling; the rich, dense Otis Vineyard Cabernet; and the smoky Red Willow Syrah.

It was just after harvest in 1984. David Lake had worked twelve hours a day, seven days a week, for over a month. No time for plans. No time for appointments. No time for much of anything other than getting grapes picked, crushed, fermented. But with the crush ended, Columbia Winery's innovative vintner had time to look ahead. He sat down with Red Willow Vineyard owner Mike Sauer, his principal grower, and discussed a tantalizing idea—planting the first Syrah grapes in Washington.

Sauer is a grape grower, not a winemaker. He supplies fruit to many outstanding wineries such as Woodinville's Columbia Winery and Walla Walla's Gramercy Cellars. Washington oenophiles make pilgrimages to the Red Willow Vineyard, one of most beautiful in Washington, located on a steep, south-facing hill with a small stone chapel at the top. In the fall, the yellowing vine leaves make this one of the most photographed vineyard sites in the state.

Conventional wisdom held Syrah wouldn't ripen in Washington and that even if it did, the first harsh winter would wipe it out. But Sauer, a shrewd man with a bit of the gambler about him, paid little heed to conventional wisdom. Unlike some growers, he didn't mind taking a chance on a new varietal if it could make great wine. After mulling it over, Sauer decided to do it. That spring he planted three acres. Not only did it survive the winter, but it also turned out stunning wine—rich, smoky, full bodied—reminiscent of the storied Hermitage wines of the northern Rhône in France.

It was like lighting a fuse on a rocket. From that beginning, Syrah exploded in quantity and quality across Washington State. The cultivation of Syrah is still in its infancy, but the buzz has already started. Gerald Boyd, wine writer for the *San Francisco Chronicle*, pegged Yakima Valley and the Rhône Valley as "ground zero for exciting Syrah." This may surprise consumers, but it didn't surprise Lake.

"Our climate in eastern Washington is very similar to the northern Rhône," said Lake, a puckish, thoughtful bearded man who died in 2009 and was one of leaders of the state's wine industry. "The Rhône is divided into two parts. The southern Rhône is more Mediterranean. In the northern Rhône—Cornas, Saint-Joseph, Hermitage, Côte Rôtie—the winters get severe. It's a continental climate, with little influence from the ocean, like Eastern Washington."

While working in the British wine trade in the early 1970s, Lake visited the northern Rhône and got his first real taste of the region's wines. He came to love Syrah, particularly the gamy, elegant version found in the northern Rhône appellations such as Hermitage. He followed that model at Red Willow. On a trip to France in 1991, Lake gave a bottle of Red Willow Syrah to Marcel Guigal, one of the finest producers in the northern Rhône.

"It's like Hermitage," said Guigal.

"That's exactly what we were hoping," replied Lake, delighted.

Such conversations convinced Lake that he was on the right track. Red Willow represented a kind of homage to Hermitage, from the rich, gamy, roasted flavor of the wines to the stone chapel Sauer erected at the top of the vineyard that was modeled after the famous chapel overlooking Hermitage. From the crucible of Red Willow, the cult of Syrah spread across the state.

Stopping in at Woodinville's Columbia Winery, I always toast to the memory of the late, great David Lake, Columbia's winemaker of many years, who helped nurture the state into a growing international powerhouse. His spirit lives on in the energy and innovation that characterize the state's industry today.

Walla Walla

Grapes, like people generally, have to suffer to achieve their potential. Too much ease, too much warmth, too much comfort—all lead even distinguished varieties to deteriorate into something only fit to drink out of a paper bag while sitting under a freeway overpass.

To make great wine, grapes need to undergo stress and strain. Like characters in Dostoevsky novels, they must overcome formidable obstacles before they can acquire their full depth and complexity, and so accomplish their mission in life, which is basically to be squished into juice and poured into a large wooden vat—a suitably existentialist end.

Among wine regions of the Northwest, Walla Walla stands out as subjecting its vines to a lot of suffering. Summer temperatures soar into the hundreds, and the long hot days and cool nights concentrate the flavors of Cabernet, Merlot, and Chardonnay grapes to astonishing levels of intensity. In winter the mercury plunges below zero, making it tough for the vines to survive. These extremities of climate cause headaches for growers, who lose plants to cold, but they bring out the best in the grapes. As a result, the valley—once known for its wheat, onions, prison, and Whitman Mission—is fast gaining recognition as the home of some of the finest wines in the world.

Some of the founding fathers of the state's wine industry call Walla Walla home, including Rick Small of Woodward Canyon Winery, Gary Figgins of Leonetti Cellar, Norm McKibben of Pepper Bridge, and outstanding newcomers such as Waterbrook Winery. All have made important contributions to the success of region, but Woodward Canyon and Leonetti Cellar have set the standard, producing wines of uncommon depth, complexity, and concentration. My apprenticeship would be incomplete without visiting them.

At One with the Fruit: Rick Small of Woodward Canyon

Woodward Canyon Winery is the logical starting point for a tour of the area. Located in Lowden, right off Highway 12 as you approach Walla Walla from the west, this winery resembles a working farm and barn rather than an upscale chateau in the California or Bordeaux style, albeit with a tasteful contemporary feel. Pickup trucks fill the gravel parking lot outside the tasting room, which is tucked inside an upscale, barnlike building at the back of the premises.

The unpretentious appearance of the winery reflects the personality and background of the owner, Rick Small, who grew up on a local wheat farm. He started the winery in 1981, and within six years, the influential *Wine Spectator* magazine had rated his Cabernet among the ten best red wines in the world. This rating catapulted Small's business into the big time. Since then Woodward Canyon has gained renown for ripe, full-bodied Cabernet and Merlot, as well as intense, barrel-fermented Chardonnay.

Despite the many awards he has garnered, Small is not the sort to rest on his reputation. On the hot fall day that I visit, he wears a pink Ralph Lauren polo shirt, blue shorts, and scuffed white tennis shoes, but he's no gentleman farmer; he had just spent the morning out in the vineyard, directing the pruning, watering, and care of the grapes. Small is a hyperactive ramrod of a guy who served as a drill sergeant in the U.S. Army Reserve. After giving workers their marching orders for the day, he gives me a tour of the winery, speaking enthusiastically about the wine-making potential of the region.

"We're starting to work with more of our own fruit," he says, showing me around the winery. "We have about ten acres planted and we're in the process of planting another eight next year. I still want to buy grapes from top-notch growers including Seven Hills, but I also want to grow more fruit of my own."

By cultivating his own grapes, Small can exert more control over the quality of the product. Through leaf stripping, cluster thinning, and careful site selection, he hopes to achieve the ideal conditions for the grapes. A fanatic for quality, Small ripped out one of his renowned Chardonnay vineyards because it was not in the right place for him. "I think this will take generations to figure out," he says matter-of-factly.

This involvement with the vineyard helped him as a winemaker. In the early years, he learned by doing. "You don't have to study wine making," he says. "I've seen no correlation between academic training and how successful you are as a winemaker. I think it's more important to pay attention to detail."

He had no recipes or formulas by which he crafted his wines. Instead, he watched over the grapes, getting to know every nuance and subtlety of their flavor, every idiosyncrasy of their growth and development. This intimate knowledge of the fruit in its various stages allowed him to make the right decisions as a winemaker.

He also learned from his friend Gary Figgins of Leonetti Cellar. "Gary Figgins and I are very good friends. We were drill sergeants together in the army reserve. We drank wine together and made wine together. Gary set an incredibly high standard for wine quality. It was immediately noticed by good wine consumers all over the country but especially in the Northwest. If you were going to make wine in Walla Walla, you had to make good wine. You had to work at it and study it and learn it. I attribute that to Gary."

As Woodward Canyon grew, Small eventually delegated the wine-making duties to Kevin Mott in 2003, so he could manage the overall operation. None of this changed the philosophy or overall direction of the winery.

"What makes Woodward Canyon wines unique is the way we do it," Small says. "We all work hard. We work very long hours doing the crush. We have a strong regimen for topping wines. We try to do everything we can to make the wine as good as it can be. We'd like to try to have that development of the wine be more on its own. We don't try to manipulate the wine too much.

"People ask me questions about why I do something. I say, 'When the wine tells me to.' I hate to sound like some airhead, but there's a lot of feeling to it. You just kind of know. It's a sense. We don't have a formula for doing anything. We keep records, but every vintage is different, just as each child is different. You're going to raise each child differently because they're different. You're going to play into their strong suits.

"I have a daughter and a son who are as different as they can be. They have different directions to go. It would be foolish for me and my wife to try to channel them both the same way. And the same thing if you're making wine."

During harvest, he shuttles between the winery and the vineyards. "You need to go there as often as you can to remember where it all happens and where the wine came from. You can't do that if you just sit on your ass down here and write checks and go to wine tastings."

As we finish the tour, he opens several bottles of Cabernet, Chardonnay, and blends. All are intense and superb.

"I'm trying to be at one with the fruit," he says as he swirls the wine in his glass. "I don't try to make big, extracted, tannic wines out of delicate fruit. I try to let the grape and the vineyard site tell me how I should make that grape into wine. That's why I was just out in the vineyard. I need to go there as often as I can to remember where it all happens and where the wine came from. My biggest plan for the future is to plant more vineyards and really get to know my own land better. I focus on that."

Gary Figgins of Leonetti Cellar

While Small concentrates on growing grapes, his counterpart Gary Figgins at Leonetti Cellar spends much of his time blending and experimenting with oak at his beautiful brownstone winery in the foothills of Walla Walla. Figgins is a laid-back, whimsical character who leaves an excellent imitation of *Saturday Night Live*'s Father Guido Sarducci on his phone answering machine: "We're alla sold outta da vino, so why dontcha call back later."

When I drive up, he's working on his Cobra sports car. Figgins wears jeans, a brown T-shirt, and cowboy boots, but despite his informal appearance, he's a virtuoso among winemakers. His bottles consistently rank among the finest in the world.

Established in 1977 as the first bonded *vinifera* winery in Walla Walla since Prohibition, Leonetti is named after Figgins's grandparents, Frank and Rose Leonetti, Calabrian immigrants who introduced him to home-made wine. These early experiences with vino led him to plant Cabernet and Riesling grapes on his property in 1974 and make wine. In 1981 *Wine & Spirits* magazine rated the 1978 Leonetti Cabernet as the best in the nation. The winery's reputation skyrocketed. As a result, the price of its Cabernet and Merlot has risen to over $100 a bottle—when you can buy it.

On a fall visit, Figgins welcomes me to his winery and switches on the light as we enter the cool, damp cellar, revealing rack upon rack of oak barrels. I'm delighted to have a chance to taste with Figgins. I've sampled his Cabernet and Merlot before and loved them both, but they're hard to find and expensive. Tasting them here with the sorcerer himself is a thrill.

At a time when many producers are getting away from using oak, Figgins is getting further into it, searching for just the right combination of wood and wine. He is a master of his medium, choosing a specific variety—such as American oak from the Eastern Seaboard, French oak, Oregon oak, or Hungarian oak—to add a particular note or nuance to the wine.

"I've been experimenting," he says, sounding like a medieval alchemist. "Wood adds body and spice to wine. The more substantial the wine, the more wood it'll absorb." Since his wines have incredible concentration and depth of flavor, they stand up well to generous oaking. Figgins moves from barrel to barrel, using a stainless steel wine thief to pour a splash for me and himself while pointing out each variety's qualities and how the oak enhances them.

"American oak accents the black fruit flavors in the wine," he says. "French oak adds astringency."

He dips the thief into a barrel of Cabernet and pours a splash for me and himself. Then he swirls the wine around in his glass, sniffs its spicy bouquet, and sips it meditatively. "Distinctive," he says.

Then he dips into the latest vintage of Merlot with his stainless steel wine thief and pours a splash for me and himself. A coffee scent wafts up from my glass. He adds it to the Cabernet. The Merlot rounds out the flavor but masks the complex spiciness of the Cabernet's nose.

"I like it the way it is," he says finally of the Cabernet.

The wines are all rich, concentrated, and spicy, with a gorgeous mouthfeel and finish. I could taste here all day, but Figgins has finished his experiments for the day.

"We don't want to be glassy eyed at dinner," he says, emerging from the cellar with a plan for what he needs to do before bottling. With his mastery of oak and instinct for blending, Figgins ensures that the Walla Walla grapes will not suffer in vain and will make possible the creation of some of the best red wines in the world.

Chris Figgins of Figgins Family Wine Estates: The Next Generation

Riding his sleek, carbon fiber road bike along the back roads of Walla Walla, Chris Figgins of Figgins Family Wine Estates looks like a competitor in the Tour de France. Focusing on his breathing and the cadence of his peddling, he follows the asphalt road as it winds through rolling hills planted with wheat, peas, alfalfa, apples, and wine grapes, with the hazy Blue Mountains rising in the distance.

I peddle hard to keep up with him, enjoying the scenery and the banter as he points out features of the landscape. As he passes a draw, he notices cold air seeping out of it, giving him real-time data about the microclimate and putting him in tune with that place.

"I'm constantly learning more about the Walla Walla Valley and how air moves around in it," he says. "From the saddle, I get a much better feel for temperature, air flow, and slope than I would from behind a pane of glass in a pickup truck."

Figgins's attention to the nuances of the local landscape has proven critical in the success of FIGGINS, an estate-grown, single-vineyard Bordeaux varietal blend from Walla Walla first released in 2008. As the son of legendary Leonetti Cellar owner Gary Figgins, Chris could have followed in his father's footsteps by focusing only on Leonetti, the cult wine that helped put Washington State on the nation's wine map. Chris joined Leonetti in 1996 and became the head winemaker in 2001. While carrying on the legacy of Leonetti, Chris also pursued dreams of his own. When a fifty-five-acre parcel of nearby farmland came up for sale, he jumped at the chance to buy it.

"It was a dream vineyard site," says the energetic, upbeat, entrepreneurial Figgins, who seems to relish discovering new possibilities in wine making. "I wanted to do a single-vineyard blend. The greatest wines in the world are identified by a specific vineyard, such as La Tâche in Burgundy, France, or the wines of Bordeaux. I told my parents, 'I gotta have it.' They were lukewarm. 'We've got all the fruit we need,' they said. 'We need a break.' But then they cautiously gave me their blessing."

For Leonetti, Chris chooses grapes from select vineyards around Walla Walla. This gives him great flexibility in putting together an ideal blend. With FIGGINS, he has limited himself to making wine from one vineyard, which has unique advantages and challenges.

"There's not a lot of room for mistakes," he says. "You have all your eggs in one basket. I have to work with what the vineyard gives. I have to be a hypervigilant farmer. It was nerve-racking at first, but after the first release, we've developed a signature and a following."

When he needs to take a break and clear his mind, he hops on his bike, clicks into the pedals, and heads out into the country, the gears meshing, the tires humming, and the road winding past vineyards toward the Blue Mountains shimmering in the distance.

Pepper Bridge Winery: History Lessons

When I arrive at Pepper Bridge, Jean-François Pellet and managing partner Norm McKibben are enjoying a glass of their rosé on the deck. The winery is located on a small hill amid the vineyards with the snow-clad Blue Mountains in the distance. They invite me to pull up a chair.

With his mane of gray hair and craggy, rugged face, McKibben looks the part of one of the godfathers of the state's wine industry, having planted some of Walla Walla's iconic vineyards such as Pepper Bridge and Seven Hills along with partners Bob Rupar, Tom Waliser, and Kent Waliser. These properties provided fruit for Leonetti Cellar, Woodward Canyon Winery, L'Ecole No. 41, and many others, putting the region on the national map.

As we sip the refreshing rosé, McKibben tells the story of Pepper Bridge.

"After a career in the construction industry, I didn't want to retire," he says. "So I moved to Walla Walla to farm apples."

"Norm's not very good at sitting around," says the dark-haired, animated Pellet.

McKibben quickly realized the region's great potential for growing quality wine grapes. In 1991 he began planting Cabernet Sauvignon and Merlot on land adjoining their apples. This was the beginning of Pepper Bridge Vineyard. After stints consulting with Canoe Ridge Vineyard, Hogue Cellars, Leonetti Cellar, Woodward Canyon, L'Ecole N° 41, and Andrew Will Winery—a who's who of state wineries—he opened his own operation in 1997.

"But I needed a winemaker," he says. "So I recruited Jean-François."

Born and raised in Switzerland, Pellet is a third-generation wine grower who always knew what he wanted to do with his life—make world-class wine. After obtaining degrees in wine making in Switzerland, he traveled the world, working in Switzerland, Spain, and Germany. His reputation for care and meticulousness brought him to the attention of Heitz Cellar in the Napa Valley, where he worked for four years before McKibben recruited him in 1999.

"I'm a good salesman," McKibben laughs.

McKibben is also a walking encyclopedia of the state's wine history, which he relates as we tour the winery. He has collaborated with noted winemakers Rick Small, Gary Figgins, Chris Camarda, and John Abbott, helping transform the Walla Walla and the state's wine industry. In between stories, he explains how he and Pellet carefully designed their facility to treat the fruit gently, from the gravity-fed fermentation tanks to the sustainable vineyard practices.

"I want to leave better ground than I got," McKibben says. "We believe in sustainability."

Outside the winery, a hawk soars overhead. Quail dart from its shadow. Frogs croak in the distance, conducting their nightly chorale. After locking up the winery for the evening, Pellet and McKibben invite me to dinner, where we can taste their wines with food— the ultimate test.

Waterbrook Reserve Wines: Crème de la Crème

They're the best of the best. Reserve wines showcase the best fruit from the best vineyards treated with the ultimate care. They run a gauntlet of tasting trials, barrel regimes, and careful analysis before earning reserve status. Winemakers select them for their quality, their varietal character, and that indescribable something that makes them irresistible.

It's always exciting to taste reserve bottles as they reveal cutting-edge wine making and the unique and evolving character of a region's fruit. For this reason, I eagerly anticipated sampling Waterbrook Winery's reserve wines at its lovely tasting room just west of Walla Walla.

"It's tremendously exciting to make these wines," says assistant winemaker Haydn Mouat (now at Canoe Ridge) as he uncorks the 2010 reserve Chardonnay, Merlot, Malbec, and Cabernet. Each wine displays its own special perfume and taste: the Chardonnay rich and redolent of pears, the Merlot ripe with raspberries, the Malbec bursting with black cherry tang, and the Cabernet brooding with chocolate and black olive flavors.

"It's a wee bit indulgent to make them," he says, swirling the chardonnay in his glass. "We ferment individual lots from the vineyards to have more control. All the blocks are kept separate. We place the wine in different oak barrels to see what happens. Everything is treated very carefully."

Such attention to detail has not gone unnoticed. Waterbrook wines have won numerous national awards, including the Wine Spectator's Top 100 Wines award for its 2009 Merlot, 2007 Merlot, and 2006 Cabernet. These impressive scores reflect not only Waterbrook's devotion to quality but also the outstanding fruit from Eastern Washington. National and international publications have increasingly remarked on the top-notch quality of Washington wine, fueling further growth of the state's wine industry. Since 2000 the number of wineries has grown from 163 to over 1,000 today.

After visiting Walla Walla, I understand the diversity of wine in the state, with so many outstanding winemakers and growers adding to the depth and quality of the industry. I learned from Chris Figgins about the limits and possibilities of focusing on one vineyard, which I've done with Ciel du Cheval, and on one varietal, Syrah, to eliminate as many variables as possible to make the best wine. Gary Figgins made a strong case for the judicious use of oak or wood. The abuses of wood are common—leading to wines that taste and smell like the inside of a barrel—but the prudent use of wood enhances wine, especially red wine. From Rick Small, I learned about understanding the vineyard, appreciating the lessons it has to teach me. From Norm McKibben and Jean-François, I gleaned some of the history of the industry and the importance of sustainability. Finally, Waterbrook inspired me to keep raising the bar.

All of it is food for thought for my own wine making. I've got a lot to consider on my drive back to Seattle.

4

Betz Family Winery

The forklift speeds a huge plastic tote to the crusher, twirls the five-by-five-by-five-foot tote like a toy, and dumps the finest Merlot grapes into the gleaming stainless steel maw of the crushing machine. Whirring gears separate the stems from the grapes, which bounce down the sorting table.

Now it's up to me and seven others to remove any stray stems, or "jacks"; underripe berries; insects; rocks; or other impurities that could mar the rich taste and elegant fragrance of Betz Family Winery's blends. I rivet my attention on the grapes gliding past on the sorting belt on their way to the crusher, not wanting to let anything contaminate the wine. My fingers fly through the fruit, trying to keep up.

It's my first day volunteering at Betz, located in Woodinville, a forty-five-minute drive from downtown Seattle. As a wine writer, I've finagled my way onto the volunteer list at Betz, a list so exclusive it's rumored someone has to die before a slot opens. I find in casual conversation the others are anything but contract laborers; they run restaurants, luxury hotels, IT departments for IBM and Microsoft. They've carved time out of their busy schedules to learn about wine from the legendary Bob Betz, one of a handful of Masters of Wine in the United States. Bob takes an occasional break from the million other things he's doing to sort grapes and answer questions.

It's a sunny October morning with temperatures in the mid-seventies. The weather is warm but not hot, perfect for wine making. I wear shorts, hiking boots, and a blue Italia T-shirt, hoping Bob will appreciate the international flare and functionality of my wardrobe. Bob doesn't seem to notice. He is everywhere at once, helping with the sorting, answering the phone, greeting people who drive up, explaining procedures to the workers.

The winery buzzes with activity this morning. Forklifts whisk totes back and forth across the concrete driveway. Workers bustle back and forth, washing bins, measuring sulfur, inoculating already crushed and sorted grapes with yeast. Everyone seems to have a spring in their step as they go about doing their work.

Crush is an exciting, exhausting time when winemakers rush to process all of their fruit in a short window. Normally, it's a six-week window at Betz. This year, because of the cool spring and hot summer, the window has narrowed to *four weeks*. Production is frenetic but not frenzied or chaotic because Bob is at the helm. An energetic man in his sixties, he projects confidence and competence. As a Master of Wine, he's a walking, talking encyclopedia of the subject, but he never comes across as condescending; he reveals his knowledge in the careful, exacting treatment of the fruit at every stage of the process. This is precisely what I'm hoping to learn from him. He oversees the harvest not as a tyrannical boss but as an enthusiastic detail freak, getting everyone caught up in the excitement of the crush.

Today he's always in motion, a blur in a work shirt, black pile vest, jeans, and hiking shoes. Though his hair and beard are gray, his brown eyes radiate energy. He seems as excited as the fruit flies buzzing around him.

Though he doesn't mention it, a lot of pressure is on him to bring in the fruit correctly and to treat it with utmost care. His business depends on getting things right every year; hundreds of thousands of dollars are at stake. He doesn't have a huge corporation backing him; he is always walking a financial tight rope. So beneath the calm exterior, Bob possesses a steely resolve. So much of wine making boils down to one's knowledge, care, and appetite for plain hard work. Bob has all this in abundance.

When I ask too many questions, he erupts, "I can't talk right now!" And he runs off. A few minutes later, he returns and answers, despite the pressures of the crush. I don't want to be a pest, but I'm dying to learn more about his approach to wine making.

His winery is located on a hill above the green fields of Woodinville, the unofficial capital of the Washington wine industry. Betz sources his grapes from eastern Washington and trucks them in for processing. The winery is small by Woodinville standards, putting out 3,400 cases of wine a year. Just down the road his former employer, Chateau Ste. Michelle, produces 3.3 million cases of wine a year. What Betz lacks in quantity it makes up for in quality, consistently scoring among the top wineries in the world.

Most important, it produces a beautifully balanced wine with great fruit and lively acidity, more old-world than new world in its style, which is exactly the signature I'm looking for in my own wine making. Traveling regularly to Europe, I've come to appreciate the balance and subtlety of Continental wines that are made to pair with meals rather than to earn monster scores from wine critics. I have no illusions about competing with Betz in wine making, but I'm hoping to pick up some of his secrets. It would be hard to find a better mentor anywhere.

Unlike the large neoclassical buildings and manicured lawns that surround Chateau Ste. Michelle, Betz Family Winery is intimate. The brown stucco entranceway frames huge carved wooden doors, reminiscent of those in Tuscany, giving the place a feel of warmth and personality, just like its wines. Yet when I step inside, it's all business: cement floors, steel fermenting tanks, white-washed walls, and plastic bins full of fermenting wine. When a worker bustles by carrying a beaker, I press against the wall to get out of the way. Everyone seems on task today.

Bob takes a break to give me the safety lecture (translated: "Stay out of the way; don't get injured") and leads me over to a bin of Merlot. "Good acidity and nice bright flavors," he says. "We like the seeds to be brown and the pulp surrounding them to easily slip away—signs of ripened fruit. There are all kinds of things we're looking for."

I listen carefully, trying to pick up everything I can. Bob leads me over to the sorting line, where we start the culling.

Ken, a muscular volunteer in khaki pants, rubber boots, and bare chest, pushes a big red button, and the sorter hums to life. Bob steps up to the sorter and explains how to cull the underripe or overripe berries, leaves, stems, stink bugs—anything that would mar the purity of the fruit. He picks through the grapes, his hands working quickly.

"This fruit is very clean," he says. "There are very few leaves." His hands shuttle back and forth across the sorting table as if they have minds of their own. I can tell he's been an educator as his explanations are lucid, pointed, and colorful. "Wine making is about pleasure," he says. "That's my philosophy."

I want to ask him what he means by that, but he dashes off to another task. Save that question for another day. The belt passes in front of us. Everyone seems to know what they're doing except me. I'm the rookie, the newbie.

I hang back, watching what everyone else is doing before I jump in. Even though I know a lot about wine, I have a lot to learn about Betz's process.

Everyone has staked out a spot on the line. People are nice but territorial. The experienced sorters have gravitated to the front of the sorter where they get the first crack at culling the fruit. When I try to take one of the coveted spots, I get butt bumped out of the way. So I return to the last spot on the sorting table, right before the grapes cascade into the tote.

The conveyor belt keeps humming. I focus on the grapes, gathering up the jacks, green berries, and the occasional bug. The berries are cool, sticky, and shiny purple like exotic jewels. The conveyor belt speeds up. Watching the grapes whiz by, I get dizzy, disoriented, and step back from the line.

"Don't vomit," Ken says. "That doesn't improve the flavor."

I take a break and get a drink of water. Vomiting wouldn't add to the quality of the wine or help my chances to continue to volunteer. I look up to see Marty, a volunteer, mixing a solution into a tall glass pipet.

"What are you doing there?" I ask.

"Don't talk to me," Marty says. "I'm doing some acid. It's very complicated."

Acid? I want to know more but decide to stick to sorting grapes. Right now, that seems the limit of my abilities. After finishing, she tells me she is adding to the grapes tartaric acid, which sometimes is used to balance the chemical composition of the fruit. I've never done this with the Ciel du Cheval Vineyard fruit I use to make my wine, but if Bob's doing it, I'll look into it. Returning to the line, I keep sorting. Soon the Ciel du Cheval Merlot is crushed and sorted.

"Let's take a break and do a *nose blow*," Bob says. Ken and the others dismantle the stemmer/crusher and hose it out. Everyone takes a break to drink a cup of coffee and to wash the sticky grape juice off their fingers.

Next, the forklift brings a bin of Red Willow fruit to the de-stemmer. Red Willow Vineyard is one of the finest sites in the state, a steep, rocky vineyard outside Yakima that resembles the famous Hermitage vineyards in France's northern Rhône region. You can be the best winemaker in the world, but without great fruit, you'll never make great wine. Bob clearly understands this.

Ken restarts the de-stemmer and sorter. The grapes tumble down the sorting table. We get back to work.

"Today," Bob says, "we're crushing grapes from two of the greatest grape growers: Jim Holmes of Ciel du Cheval and Mike Sauer of Red Willow Vineyard."

I appreciate Bob's take on the vineyards. In working for Chateau Ste. Michelle, he got to know the best vineyards in the state—particularly, their quality, stylistics, and character, and the personalities of the people who run them. Most winemakers know a few vineyards well; Bob got to know most of them well. He obviously zeroed in on the primo ones.

A prosperous-looking couple arrives in front of the winery. The woman wears a red jacket and pleated skirt; the man wears a blue blazer, khaki pants, and tasseled loafers and carries what looks like a very expensive leather briefcase. They look completely out of place; everyone else wears old T-shirts, jeans, and hiking boots.

"We'd like to talk to you about life insurance," the woman says, smiling as if this is exactly what Bob is dying to hear. She is totally oblivious to the activity going on around her. I'm thinking, *If they don't leave soon they will* need *life insurance.*

Bob doesn't lose his composure, even though he's right in the midst of one of the busiest harvests ever. The phone rings. "I've got to take that," Bob says, rushing off. Duane, his assistant, ushers the couple off the property.

We break at lunch, having crushed and sorted the Ciel du Cheval and Red Willow fruit. Now Cathy Betz, Bob's wife, brings out vegetarian lasagna and sandwiches. Cathy is a petite, dark-haired woman with a calmness that complements Bob's restless energy. Several of the volunteers have brought wines hidden in paper bags to try at lunch. Part of the Betz wine-making apprenticeship involves trying wines from around the world, analyzing them for their smell and taste, and trying to guess their origin.

Steve Wells, a volunteer, opens the first bottle. I take a taste along with the others. Bob asks first whether it's old world or new world. I guess new world. Wrong! It's old world. What's the varietal? I guess Merlot. Wrong! It's Barbera. What year? I don't venture to guess. Someone else guesses 2005. It's 2003.

I think of myself as pretty knowledgeable about wine, but I'm going to have to step it up. Blind tasting is a way to hone your skills as a taster and smeller, making you a better winemaker. I vow to drink more wine. I'll make it my New Year's resolution!

Bob opens a second bottle and pours it into everyone's glasses. "Old world or new world?"

I inhale deeply and detect the ripe, plummy fruit of a new-world wine. "New world."

"That's correct," Bob says. "What region?"

I take another sip. "California."

"That's correct," Bob says. "What year?"

I take another sip but can't decide. I'm stumped but pleased; I got two questions right. Finally, Bob steps in. "It's an older wine," he says. "Ten years old. You can tell by the character of the fruit. Anyone know the producer?"

Silence.

"It's Villa Mt. Eden," he says, removing the brown paper wrapper. I feel vindicated; I didn't disgrace myself. There's another bottle to taste, but I need to get back home to finish up a story. Bob insists I take a bottle of his Père de Famille Cabernet for helping out; he mentions in passing that he believes that it's the finest wine that he's made. I thank him profusely, hoping to get invited back.

A week goes by. The harvest is in full swing, but I haven't heard from Bob. Did I say or do something wrong? Did I ask too many questions? I felt as if I had just scratched the surface of Bob's knowledge. Sorting may be a menial task, but watching him make decisions is a real education.

I'm tempted to email him, but I don't want to bug him. Everyone is crazy busy during the harvest. Everything takes a back seat to the grapes. I know this myself. I've been bringing in my own grapes for years. Even with a small operation, its coordination is very complicated. I tell myself not to worry, but I continue my hand-wringing. Then on Monday, I receive an email from Bob.

> Hi Folks,
>
> A big week of crush scheduled for next week, so we're sending out a call for volunteers. Just mark the day(s) below that you are interested in joining us and I'll get back with you specifically to confirm. Some

days require more than others so we'll verify with you specifically on the days. 9 am start all days. If you can't make it in, no need to reply.

Thanks so much, in advance.
Bob

Yes! I sign up for Tuesday, eager to learn more.

Tuesday dawns bright and sunny, and the poplar trees on the road to the winery are turning yellow. Duane, an unofficial cellar master and jack-of-all-trades at the winery, greets me on his way inside. He's lively, upbeat, always in motion. No one is sitting around.

The volunteers are already in their place, ready to sort the grapes. Bob has a deep bench. There are a number of new faces, many of them from the restaurant and beverage industry. The pace is faster today, the volume larger.

I bat clean-up again, this time with Jeff Nouwens, whom I recognize from the last sorting session. "Sorting is a menial task, but the conversations are mind-challenging," he says, picking through the fruit. "You get to stand at a sorting table with Bob Betz and ask him questions. It's like going to school. It's a free education from one of the best winemakers anywhere."

I nod in agreement, telling him about what I picked up from my last day here. As we're talking, a stem drops into the fermenting tote. Nick Highland, a senior volunteer, plucks it out.

"I'll take the heat," Jeff says. "I only have one good eye."

Jeff smiles wryly. He's obviously used this line before. He's been volunteering at Betz and other wineries after retiring from PACCAR. "I got in here through the AARP," he jokes.

Like the other sorters, he's ridiculously overqualified for the job, but he enjoys the chance to socialize, share the excitement of the harvest, and learn about wine making.

Bob steps to the line and fills us in about the Alder Ridge Vineyard fruit we're sorting.

"I'm so pleased," he says, his fingers flying through it. "It's a little cold, so that makes fermentation a little harder, but it's nice and clean. It was a tough year, but we've tortured the vineyard people into growing it the way we want. I've been buying fruit from Alder Ridge Vineyard for eleven years.

We get an acre of Cabernet from them, three tons per acre. It took a number of vintages to optimize the site. The growers have gotten religion. They've got real attention to detail. They really stick to it."

Bob knows exactly what he wants from a vineyard and then works with the growers to get it. He is one of the leaders in making Washington wine among the best in the world.

"Alder Ridge overlooks the Columbia River," he continues. "The down-side is the wind from the west. It can heat up the vines. I had them just pick Merlot from the east side, so what I see here is pure fruit with only a little sunburn." His hands work back and forth through the grapes. "The beauty of Cabernet is that it's a grape that takes a licking and keeps on ticking. It's a thick, robust berry. Stomp me, crush me, make me whine!"

Then he's off to complete another task. The volunteers resume their chatter. Therese Andre, a project manager at IBM, explains why she's here. "I just got bit by the wine bug," she says. "I love the complexity, all the decisions you have to make. The more you know, the more you realize you don't know."

This remark makes me feel better about my lack of knowledge about the Betz operation.

Quentin Incao, director of operations for MTM Luxury Lodging, which owns the Willows Lodge in Woodinville, explains why he's joined the sorting line. "I love Betz wines," he says. "So much goes into it. Working here is like taking a course from A to Z."

Everyone resumes sorting, seemingly impressed by these testimonials. I shove my hands into the grapes, which are cool to the touch, squishy, sticky with sugar. Working on the sorting line you really get to feel the texture of the fruit. I shove a few grapes in my mouth; they have a deep, rich blackberry taste, softer, and seemingly less tannic than grapes from other vineyards. I prefer grapes with more structure or tannin, but the taste is delicious; they will likely be used as a part of a blend.

When we take a break, I go inside the winery for a cup of coffee. I run into assistant winemaker Katherine House (now operating the House of Wine in Idaho). Affectionately known as "Kat," she does chemical analyses—running tests on the sugars, acid, pH—checks berry size and cluster size, and seeks the ideal fermentation techniques for each lot. In addition to analyzing the data, she also tastes the fruit.

"It's like Emile Peynaud used to say," she says, referring to the father of modern oenology, "'I take all the numbers and put them in a drawer because the fruit itself tells you what to do.'"

"How do you know what the fruit is telling you?" I ask. I've been relying on numbers to make my wine-making decisions. Without them I'd be flying blind, but there's obviously more to it than it.

She replies that she listens to Bob. "He's seen so many different things," she says. "He knows what flavors he's looking for and how to get them."

During the harvest, she works 7:00 a.m. to 6:00 p.m., five or six days a week and sometimes more. The long hours are challenging, but she relishes learning from Bob, who has some fifty years of harvests in Washington under his belt. Every harvest is different. This one is no exception. "It's like a box of chocolates," she says. "You never know what you're going to get until you open it."

When the sorter starts up again, Bob returns to the sorting line. I ask him about this year's harvest.

"It's one of the strangest on record," he says. "It probably has highest heat of any vintage in the last decade."

In the spring, the heat accumulation at Red Willow Vineyard was the third lowest on record. Then in the summer, the heat approached the highest on record. Such temperature extremes pose a challenge to a winemaker.

"The sugar was high when we harvested, but we haven't got overripe fruit," he says. "It shows how much we've learned."

"Hey, a rock!" someone shouts, picking it out of the grapes.

"It will add texture to the wine," Bob says, not missing a beat. "A mineral note. It's just like the description for La Serenne [one of his Syrahs]—liquid rock."

After we finish the batch of grapes, the sorting table continues to run. Jeff pushes the tote full of crushed grapes into the winery. The others grab a cup of coffee. I'm left standing at the sorter, wondering if someone is going to turn it off. I look down at the big red button. Should I push it? I've seen others do this, but they were more experienced than I am. I don't want to do something wrong, yet it seems silly to keep the machine going with no grapes to sort. I reach down and push the button. The machine slows to a halt. No one reprimands me. It may not be much, but it feels like a breakthrough.

Things are buzzing at Betz Family Winery when I arrive on a Sunday in mid-October. A forklift whizzes back and forth across the concrete crush pad, ferrying plastic totes full of grapes to the stemmer/crusher. I park my car and walk toward it. The forklift screeches to a halt, and winemaker Louis Skinner gets out to introduce himself.

"I gave the volunteers the day off," he says. "Everyone was so spent."

Louis is tall, with glasses, pomaded hair, and an intensity matching that of Betz's founder, Bob Betz (who sold the winery in 2011 to Steve and Bridgit Griessel but remains the consulting winemaker). Louis rose this morning at four o'clock and must be running on fumes, but the excitement of the harvest overwhelms everything else.

He gives me a quick safety lecture: "Steer clear of the forklift. Give it plenty of room. You don't want it to run over your foot."

"Got it!" I say. "What can I do to help?" The last time I volunteered at Betz, there were twenty others, and I almost had to wait in line to help. Today, it's just Skinner, assistant winemaker Jonathan Villaseñor, and me. I feel fortunate to be able to work the entire stemmer/crusher, learning the intricacies of how Betz stems and crushes its fruit. Louis explains the operation. He demonstrates how to clean a screen that removes the seeds, which add tannin to the wine, something that Betz doesn't want.

"Keep your hands away from the gears," he says. "If the machine gets clogged, just push this button." He points to the big red button. "That will shut off the machine and give us a chance to clean it out."

Louis jumps back in the forklift and ferries a bin of Olsen Vineyards Petit Verdot into the hopper of the stemmer/crusher. In between ferrying bins of grapes, Louis fills me in on the harvest. "This is the second most condensed harvest we've had," he says. "We got 50 percent of our grapes in six days. Half the harvest in a week!"

The compressed harvest explains the absence of volunteers today. This allows me to get even more experience. I clean out the seed tray, dumping the seeds onto the grass.

After the machine de-stems the grapes, they head up a conveyor belt toward the crusher. Louis adjusts the rollers to treat them gently. Then it's up to me to add sulfite.

"Put in two hundred milliliters per bin," he says, pointing to a pitcher of water with sulfite in solution. "That will shock the wild yeast and allow us to introduce the cultured yeast."

I pour the sulfite solution into a measuring cylinder and dump it into the grapes.

Betz sources fruit mainly from the Yakima Valley, working with fifty different vineyards. "It's a big canvas to work with," Louis says. "It's ground zero for quality."

Logistics are complicated with getting fruit from so many sources, especially with such a narrow window this year, but Louis keeps things on track. They're nearly finished with the crush for the year.

"The harvest started late," he says. "We had moderate temperatures through the summer, and then six weeks ago, the weather cooled. There were freezes recently on Red Mountain. Some other wineries were struggling with low Brix, but we trimmed our crop load, which helped it to ripen early. We're always trying to improve."

After we finish crushing the Petit Verdot, Louis suggests I go inside the winery to join Jonathan as he tastes through the ferments. Jonathan stands below a whiteboard listing the names of Betz's vineyards. It's a game plan for the harvest, with figures listing the Brix, pH, and acidity levels—the crucial measurements of grape maturity. Sourcing from fifty vineyards means keeping track of a lot of ferments.

"Making wine is a lot of work," Jonathan says, but he doesn't seem daunted by it. In fact, he seems to relish tasting the wine at this phase of the process.

We start by tasting a Pinot Noir. "Pinot is very finicky in the vineyard and in the winery," he says. He hands me a plastic cup of Pinot Noir from Lazy River Vineyard and calls Louis over for a taste.

"It's gonna be sick!" Louis shouts, swirling the juice in his mouth.

Jonathan continues to taste through the red plastic cups with fermenting juice. He measures the Brix level and samples the liquid every couple days. "I'm trying to figure out what's ready to process depending on the time on the skins," he says. "Then we can adjust the ferment."

Tasting wine at this stage has never been easy for me. It's hard to get beyond the yeast and other flavors in the ferment and to envision what the wine will become. I listen to Jonathan describe what he's tasting and smelling, and see if I can discern it.

We try a Red Willow Syrah. "It's just beautiful," says Jonathan admiringly.

It's yeasty and tart with tannin, with a peppery nose and a fruity, smoky taste. I'm beginning to learn how wine tastes and smells at this stage of the process.

After he finishes tasting, Jonathan leaves for the day. "Hope you get some rest, man!" Louis says.

With the crush finished, Louis cracks a bottle of 2016 Père de Famille to give me an idea of what the wine will become. I swirl the wine around in my glass before I taste it. It's deep purple, devoid of sediment, with a savory bouquet and an intriguing pirouette of a finish.

"Beautiful!" I say, thanking him. I wash the grape juice off and get ready to leave. Louis continues to work.

"Get some rest!" I tell him as I leave, knowing he'll probably ignore my advice, at least until after the harvest.

As I walk back to my car, I turn all of this over in my head: the meticulous treatment of the grapes, the reliance on taste to guide decisions about picking, and the need to get the best fruit. I already have an advantage in this last regard, as I'm getting grapes from Ciel du Cheval, but I need to improve in these other respects if I ever hope to make wine anywhere near Betz's.

While I'm sure there's more to learn from sorting at Betz, I decide I want to experience the crush at DeLille Cellars, which is right up the road from Betz. Winemaker Chris Upchurch is one of the pioneers of the state's wine industry. He also sources fruit from Ciel du Cheval Vineyard and is renowned for concocting award-winning blends. What's his secret for turning Ciel fruit into delicious Bordeaux-style blends? His volunteer list is reputed to be hard to crack, but if I can get in at Betz, I might have a shot at DeLille.

5

DeLille Cellars

The Art of Blending

It happens right after the coffee break. Stepping away from the sorting table, I barely have time to choke down a roll before it's time to get back to work. I take my coffee with me as I join a dozen other volunteers to step up to the sorting table in a cavernous warehouse smelling of yeast and loud with the reverberating hum of the sorting machine.

"All the slackers—I mean, volunteers—back to the line," jokes Craig Nickel, DeLille's cellar master (now at Avennia), who oversees operations in the wine cellar.

I put my coffee cup down next to the picking bins and join the sorters to resume culling. The clusters fall onto a mini conveyor ramp that brings them to the top of the de-stemmer. After running through it, the berries bounce onto a second ramp, and we sort through them.

"Is that yours?" Nickel points to my coffee cup.

I offer to move it.

He picks it up and pours it down the drain. "No food out here!"

"Okay," I say, chastened.

After we've finished a bin of Cabernet, I start cleaning up the stray grapes, trying to help out. I grab a handful of seeds and skin right above the crushing rollers.

"Don't get your hand below that line," Nickel warns me. "No grape is worth losing a finger."

I nod again. Though I've made wine myself for years, it's challenging to adapt to a larger, more regulated system. My system is more ad hoc, as suits a garage winemaker; DeLille's system is designed for a commercial winery. I vow to pay attention and follow instructions so I will get invited back. It took several calls to get on the volunteer list at DeLille Cellars, one of the premier wineries in Washington State. I had to request a specific date several weeks in advance. As at Betz, many well-educated people are willing to work for a bottle of wine a day. Like me, they are caught up in the excitement of this ancient ritual, obsessed with this mysterious liquid.

DeLille's chateau, a pretty, French-inspired building, sits on a hillside south of Woodinville. The working winery, however, is located in a light-industrial zone north of Woodinville, a facility that's all business with concrete floors, foil insulation walls, and high ceilings to make room for the hundreds of barrels stacked in the back. This is the guts of the operation, where the real work of the winery takes place. Wine making has a romantic aura about it, but I'm finding that it's really about hard physical work, especially during the harvest.

Where Betz produces 3,400 cases per year, DeLille does ten thousand. That's a lot more than Betz but a great deal less than nearby Chateau Ste. Michelle, which produces 3.3 million cases a year. DeLille consistently ranks among the top boutique wineries in Washington. Its wines routinely earn ninety-plus scores from Robert Parker, one of the most influential critics in the wine business. Parker has dubbed DeLille one of the few five-star grand cru wineries in Washington State.

DeLille has served as a training ground for many of the best young winemakers in the state. The late Ross Mickel of Ross Andrew Winery, the late Lance Baer of Baer Winery, and others got their start here before launching their own wineries. Chris Upchurch, the executive winemaker, has a reputation for hiring aspiring winemakers and developing them until they're ready to go off on their own. If I do my job and avoid screwing up, I may have the chance to do the same. I return to the sorting table.

Winemaker Chris Peterson hops on the forklift. The lively, bearded, boisterous Peterson maneuvers the forks under one of the totes weighing a thousand pounds and lifts it up above the cluster sorting table. He deftly tilts the bin so the clusters fall down a chute and onto the sorting table.

"It's Junkyard Ciel du Cheval Cabernet," he announces to the volunteers. "It's some of the best fruit we get from Ciel." It comes from a block or section where the vineyard used to dump its old farm machinery. Now it's a vineyard in its own right, as the land on Red Mountain is so valuable for growing grapes that there's no room for junkyards.

Unlike Betz, DeLille has two sorting tables: one for clusters and one for grapes. The volunteers at the first table sort clusters for any rot or leaves, and sorters at the second table remove any stems left after the de-stemming.

"It's a double hand sort," explains Peterson. "The shaker table sorts the leaves and bad clusters. Then the conveyor belt brings them to the de-stemmer. After they're de-stemmed, there's a secondary sort for underripe berries and stem fragments. We're trying to do everything we can to make high-quality wine."

I take my place at the first shaker table, diligently plucking stray leaves from the gleaming clusters of fruit coming down the table. The table vibrates, making the clusters dance and spin, separating out anything that might taint the blend. Having worked a similar job at Betz, I think I know the drill, but I don't want to assume anything. I want to participate, observe, and gradually work my way into the operation.

As at Betz, I learn most of the folks have been volunteering here for years. They are not your typical manual laborers—doctors, lawyers, businesspeople—but enjoy being involved in making wine. After an hour, I move to the sorting table to try my hand at removing the jacks from the berries.

Eight people are at the table, four to a side. The table vibrates, making the grapes, or berries, dance along like Mexican jumping beans. Any remaining stems need to be plucked out. The stems can give wine a vegetal taste and smell that DeLille wants to avoid.

"You can stand over here," says a guy wearing a Sonics sweatshirt. "Get the stems as they show themselves."

Like the other table, the volunteers chat while they work, talking about last night's University of Washington football game or the winery mailing lists they are on. The smaller, more prestigious houses sometimes sell most of their wine to those on their mailing lists.

"Do you know Cayuse?" says Tracey, a realtor. "It's so hard to get on their mailing list. I found out about them when I was visiting my brother, who is in prison in Walla Walla. I started going to the local wineries and soon became a fan."

"Yes," I say. "I wrote a story about the owner, Christophe Baron. He's a fascinating guy. He grew up in France and recognized the potential of vineyards consisting of stones. He makes killer Syrah."

Heads nod, duly impressed with my insider knowledge.

The grapes dance down the sorting table, a white conveyor belt bringing them to the crusher. The workers stand on the sides of the table, removing the underripe berries, jacks, leaves, or the occasional stink bug. I'm at the end of the table, the last line of defense before the jacks or other impurities could plunge into the steel jaws of the crusher below me.

The machine hums, the drums of the crusher whir, and the grapes make a satisfying squishing sound as they hit the drums. Then they are pumped through a hose into the fermenting tank. The intoxicating smell of yeast and grape juice fills the air. Fruit flies circle excitedly.

After an hour, I'm sticky with grape juice. I have a hard time focusing on the table, as the monotony of the task and the smell of propane from the forklift makes me dizzy. Lunch comes just in time.

Sandwiches and soft drinks are available on a table in the office. The volunteers file out to each lunch, several of them wearing purple DeLille "Rat du Cave" or "Will Work for Wine" T-shirts. I sit down next to Peterson, who leans back in a plastic stacking chair, taking a break. He's learned to pace himself during the harvest, which usually lasts from the beginning of September to mid-October. It's a marathon, working every day until the crush is over.

"It's tiring," he says. "It's thirty to forty days in a row. There's always something to do. Your body gets used to it."

Normally, he and cellar master Nickel divide up the day-to-day duties of the wineries, with Nickel overseeing the cellar and Peterson overseeing the wine making. But Nickel hurt his shoulder in a motorcycle accident, so he's doing most of the lab work while Peterson runs the forklift and does more of the physical work.

"We're all just trying to get it done," Peterson says, rubbing his purple-stained hands together.

Peterson and Nickel make the everyday decisions while executive winemaker Chris Upchurch guides the overall strategy for the harvest: Which fruit to pick? When to pick it? How to ferment it? These decisions are critical in crafting a superb bottle of wine.

"Can I continue to help out?" I ask.

"You can come and observe," he says. "And help if we need it."

Not exactly a ringing endorsement of my efforts, but I'll take it for now.

"Fire it up!" Nickel yells. "Lunch break is over."

The crew is vigilant, staying focused on sorting the clusters and grapes. It would be hard to find a more diligent bunch even if you paid them. Making great wine inspires devotion.

At the end of the day, the volunteers begin counting the bins left to sort.

"FOUR!" We yell. The forklift dumps the grapes into the cluster sorter.

I step away from the table to take a breather. The grapes come bouncing down the table. It's back to work, culling jacks and underripe berries.

"THREE!"

I keep sorting, trying to keep focused on the grapes blurring past.

"TWO!"

The grapes keep coming. I work to keep any underripe ones from the crushers.

"ONE!" The last of the grapes come down the sorting table. My eyes are glazing over, but I keep going until the machine goes off. Done!

While the other volunteers pick up bottles of wine for helping out, I spot the executive winemaker in the office. Chris has gray, shoulder-length hair; blue eyes; and an amused smile. He wears jeans, old running shoes, and a purple Demptos Napa Cooperage T-shirt. Fruit flies follow him like old friends.

Rather than bottling single varietals such as Cabernet or Chardonnay, Upchurch prefers to blend wines, paying homage to Bordeaux-style wines with his Chaleur Estate and D2 red wine blends and Chaleur Blanc white. He sources fruit from the top vineyards of Washington, including Ciel du Cheval Vineyard.

An avid Francophile, Upchurch loves the wines of Bordeaux, using their blending practices as a model for his wine making. "I don't think you can make wine by committee," he says. "If you do, you get the least offensive homogenization. As founding winemaker, it's my responsibility to establish a style. It has to be different. That's one of the lessons I learned from the French."

Upchurch is a very skilled winemaker and a likable, easygoing character. I've met him at wine events and know something of his story. As a young man, he traveled through France and Europe, visiting wineries and educating his palate. He started out in the wine trade as a retailer, learning all the various kinds of wine, domestic and international. In 1992 he became the winemaker at DeLille Cellars, where he drew on all his experience to create Bordeaux-style blends.

He's learned his craft from the founding fathers of the Washington wine business: Bob Betz; David Lake, the late winemaker at Columbia Cellars; Walter Clore, the WSU professor of agriculture who pioneered the planting of *vinifera* grapes in Washington; Rick Small of Woodward Canyon Winery; Gary Figgins of Leonetti Cellar; and Mike Januik of Januik Winery.

"They were the ones doing the early work," Upchurch says, leaning back in an office chair. "David Lake taught me there's way more to wine making than pH and TA [titratable acidity]. It's how you *think* as a winemaker. You think about it as a craft. People talk about it as an art or science, but I look at it like a craftsman. There are a lot of artistic aspects to wine making, but I'm happy being a craftsman."

Upchurch likes making wine in Washington because everyone is so helpful. "We loan other winemakers our equipment, and they help us out," he says. "In Washington, everything needs to get better. I've stood in most of the great vineyards in the world. Our vineyards are as good as any in the world. Look at Quilceda Creek, which has the best Parker scores of any winery in the world over the last two years, with four hundreds and two ninety-nines. Now Parker scores aren't everything, but nobody has that kind of record. You have to look at Washington."

As good as the vineyards are now, Upchurch believes they can get a lot better. "The next quantum leap will take place in the vineyards," he says. "I've traveled all over the world—California, Australia, Burgundy—to learn about their growing techniques and see what I can apply here. I enjoy working with Ciel and Klipsun on Red Mountain and Boushey near Grandview. David Lake taught me you need to get your nose in the vineyard and don't take it out."

Before and during the harvest, Upchurch makes frequent trips to the vineyards to get a sense of when to pick. As the fruit gets processed, he forms an idea of how it will fit in with the future blends.

"I think science is overrated in wine making," he says. "It keeps the wine stable, but wine making is an art and craft. The blending trials are the art part. The final product should be greater than the sum of its parts."

He points to the whiteboard with the field samples from the various vineyards measuring the pH, tartaric acid, and Brix levels to compare them with those of previous years. "It's different every year," he says. "It forces you out of formulas. It forces you to listen. The better winemakers realize they have control over nothing."

While most Americans recognize wines by the varietal—Cabernet, Merlot, Chardonnay—DeLille follows the French Bordeaux model of blending varietals, such as Cabernet, Merlot, and Cabernet Franc, to make what it considers the best wine possible. Its 2015 D2 proprietary red wine consists of 58 percent Merlot, 36 percent Cabernet Sauvignon, 5 percent Cabernet Franc, and a splash of Petit Verdot combined into a rich, fruity, polished package. The result makes a strong argument for blending.

The whiteboard helps Upchurch chart the progress of the harvest and suggests how future blending might go. Despite the hectic nature of the crush, he seems to relish his job and role as one of the leaders of the state's wine industry. "At the end of the day you have something to show for your efforts," he says. "I'm very comfortable with my job. It brings a lot of pleasure to a lot of people."

The long days and the hard labor are simply part of it. "It's October," he says, raising his hands as if asking a question. "If you can't get psyched for the harvest, you shouldn't be in this business."

I thank him and get back to work, cleaning and washing the de-stemmer and sweeping the grape skins from the floor, doing the stuff of my apprenticeship. My back aches from a long day of sorting, but I'm energized by the tactile nature of the job and the enthusiasm and professionalism of the DeLille team. Upchurch's genius for blending is inspiring. I have plenty to learn about fermenting even one varietal at a time, but I've dabbled with blending Cabernet and Syrah, and liked some of the results. After I finish sweeping, I wash up but have trouble removing the purple stains from my hands—the mark of a true cellar rat.

6

Decision-Making at Canoe Ridge

You can't understand Washington wine without understanding Chateau Ste. Michelle, the state's largest producer. Ste. Michelle, with a yearly capacity of nearly 3 million cases, and sister winery 14 Hands (1 million cases) are the largest wineries in the state and among the leaders of the wine industry.

Headquartered in Woodinville, Washington, the flagship company founded in 1933 bestrides the state's wine industry like a colossus, serving as an advocate for Washington wine by encouraging a level of cooperation that has lifted the industry and allowed it to progress quickly. When a cold snap in 1996 decimated the grape harvest, Ste. Michelle stepped in to sell grapes to smaller wineries, taking the long view that a healthy industry needed a wide range of houses. Winemaker Katie Nelson and her Ste. Michelle team exemplify best practices in oenology and viticulture, while distinguished alumni such as Bob Betz of Betz Family Winery and Mike Januik of Januik Winery take the industry to new levels of excellence. The company's marketing muscle opens the door for Washington wine around the globe, selling a broad portfolio of delicious, inexpensive varietal wines and superb single-vineyard Chardonnay, Merlot, and Cabernet, as well as the high-end collaborative ventures Eroica Riesling and Col Solare. Ste. Michelle boasts a tantalizing selection at all price points.

Shortly after turning twenty-one, I purchased my first Washington wine, a bottle of Chateau Ste. Michelle Johannesburg Riesling, the wine that put Ste. Michelle on the map. In a 1974 *Los Angeles Times* blind tasting of nineteen white Rieslings, Ste. Michelle's 1972 Riesling ranked first, vaulting the winery and state's wine industry into the national spotlight. The awards have been pouring in ever since. In the following years, the company branched out to produce a variety of red wines, including Cabernet, Merlot, and Syrah.

I'm especially interested in seeing how the winemakers make red wine at their Canoe Ridge Estate facility and vineyard in the Horse Heaven Hills AVA. It can produce up to a million cases a year, which is way more than any other place I've visited. The facility is located on the Columbia River, several hours east of the main offices. The head winemaker, Brian Mackey, invites me to tour the facility during the harvest, the most exciting time to visit.

On a Friday in late September, I rise at dawn and drive east toward Paterson, Washington, where the facility is located. By the time I cross Snoqualmie Pass, the dividing line of the state, the sun rises, and the landscape comes to life. Highway I-82 follows the Yakima River, winding past orchards, vineyards, and fields golden with the harvest.

At Prosser, I head south, leaving the farm country behind and crossing a deserted landscape, with little vegetation and dusty fields descending toward the Columbia River. As I approach the river, splashes of green appear against the low brown hills—vineyards. Here, there, and then everywhere, they carpet the contours of the hillsides with green. I've entered the Horse Heaven Hills AVA where Chateau Ste. Michelle's Canoe Ridge Estate vineyard is located.

I turn right and head up the road to the Canoe Ridge facility. It dwarfs all the other places I've visited. The large white building looms above the river like an industrial facility. Parking my car, I enter the large building and ask for winemaker Brian Mackey.

Brian welcomes me into his office and offers me a seat. A friendly man wearing jeans, a blue rugby shirt, and hiking boots, he exudes enthusiasm for wine making. Wine has played an important part in his life, as he explored his grandparents' Napa Valley vineyard as a child and worked summers in high school at a family winery in Washington State. After earning an English degree from Boston University, he worked in publishing, in software, and in the film industry as a cameraman before succumbing to the siren song of the wine industry. Eventually he interned at Chateau Ste. Michelle and quickly climbed the corporate ladder.

His passion for wine making is especially evident at the harvest. "Winemakers love the harvest," he says. "Christmas has nothing on the harvest. The harvest is an extremely stimulating time. You're making the most important decisions regarding the wine. It's invigorating to be that busy and have so many balls in the air. Harvest is the key to making great wine."

Though he's in the middle of the harvest, the most hectic time of year, he has planned a full day to help me understand how the winery approaches its craft. I'll shadow him as he makes the rounds of the vineyards, deciding when to pick the grapes and how to ferment them.

"We let the grapes speak for themselves," Brian says, leaning back in his chair. "We apply the same simple principles as other wineries, but we use more people and more technology. We embrace technology if it improves the quality of the wine and our efficiency. We're using more people and equipment but a similar approach to a smaller winery. We have a balance between classic historical technique combined with logistics."

As I'll see later that day, the winery employs techniques the smaller wineries choose not to use or cannot afford to use, allowing it to produce vast quantities of quality juice. Unlike some of the smaller houses, Ste. Michelle seeks to create a house style.

"No two vintages are the same, but we're extremely consistent" with our wines, he says. "It's really a fun puzzle to tackle every year."

This year is a fairly cool vintage, which allows the grapes to retain a lot of natural acidity. The temperatures dropped to the thirties over the preceding weekend. "It's a high-quality vintage, but everyone gets stressed out," he says. "The temperatures are a little alarming. If there's frost, it does affect the fruit."

This risk is common in Washington. With its northernly latitude, the winemakers worry about frost damage. This variable is just one of several they juggle every year. Though many see wine making as a romantic activity, Brian sees it more practically.

"There's a romance of what it is to be a winemaker," he says. "A winemaker is a decision-maker. You shepherd the grapes to the bottle. The decisions you make will have a direct result in the wine."

Today Brian will illustrate the four key decision points of the harvest. Decision 1: When to pick the grapes? The key numbers—for example, the Brix, or sugar content; titratable acidity; and pH levels—help winemakers determine when to pick, but Brain argues the final decision comes down to taste.

Decision 2: What type of fermentation vessel should be used: oak barrel, stainless steel tank, or clay amphora? Each of them has unique qualities. Winemakers must match the varietal to the fermentation vessel.

Decision 3: When to press—that is, when should the wine be pressed off the skins? For white wines, it could be immediately. For red wines, it ranges from a few days to a month, depending on how quickly the yeast converts the sugar in the grapes.

Decision 4: What type of barrels? Most red wine is aged in oak barrels, but there are many options here. French oak? American oak? Hungarian oak? Should the barrels be new or neutral? Winemakers can argue for hours about these matters.

"These four decisions determine the fate and the style of the wine," Brian says. "It's the same at smaller wineries. Whether you scale up or down, you have to make these decisions. The differences here are the logistics."

To see how this works out in practice, we hop into his black SUV to make the first of these decisions—that is, when to pick the grapes. Gravel crunches under the tires as we climb the low ridge above the Columbia River. Explorers Lewis and Clark thought this area looked like an overturned canoe and named it Canoe Ridge.

As in other wine regions, geology is critical to the vineyard's success. Some 12 million years ago, lava flows sluiced across the landscape, bringing layers of basalt. As noted previously, during the last Ice Age thirteen thousand to fifteen thousand years ago, the great Missoula floods deposited a deep sandy loam with cobblestones, creating a well-drained soil with low fertility. Add warm weather, strong winds, and the moderating effect of the Columbia River, and—bingo!—you have an ideal site for grape growing.

Exiting the vehicle, we walk toward an overlook near the top of the ridge at 950 feet. The vast 559-acre estate, planted in 1991, spreads out beneath us; undulating rows of vines descend toward the Columbia River, which is heavily dammed, wide, and flat as a lake at this point. I enjoy the view. There's something soothing about seeing how a vineyard both conforms to the landscape and shapes it. Nature and culture combine into a harmonious whole.

The wind gusts, blowing dust in my face. We turn around and head toward a block that is being picked. Because the slope is gentle here, the company uses a mechanical harvester, one of the technical innovations Brian talked about. We approach the Pellenc grape harvester, which looks like something out of *Mad Max 2: The Road Warrior*. As the C-shaped tractor straddles across the top of the vine rows, with one set of tires on one side and the other set on the alternate side, a mechanism in the middle shakes the vines, de-stems the clusters, and drops the grapes into a bin below.

"There's some damage caused by the harvester," Brian says. "We might lose a cluster here or there. But it's quick and efficient, and the quality of the fruit is high."

None of the other wineries I've visited use such a machine, but none produce as much volume as Ste. Michelle. This is one of the technical innovations that allows them to produce huge quantities of consistently fine wine.

We walk farther into the vineyard to Block 25 to determine if it's ready to pick. Brian heads down the rows, randomly tasting Merlot grapes. I do the same.

"We're looking at flavors," he says. "We're looking for sugar and acid balance." Chewing on a grape, he says, "These are really dried out. They're too astringent. The tannins are not ready."

Tasting the grapes, I see what he's talking about. I like tannic wines, but these grapes make my mouth pucker.

We cross the road to Block 24, which looks identical to 25, the taut wires of the trellises festooned with a canopy of leaves and hanging clusters of grapes. Brian resumes tasting. "The tannins are chalky, but I'm getting more berries and jam," he says. "We're moving out of watermelon toward dark fruit flavors."

I chew the grapes, finding less astringency than those of the other block. I detect some of the fruits he describes and better understand his decision to pick. These grapes offer the possibility of a round, smooth wine.

"These blocks have similar chemistries," he says, chewing on a grape. "Why are they different? We don't really know. But physiologically these grapes [in Block 24] are saying, 'It's fall. I'm done. Pick me.'"

The first decision has been made. He puts Block 24 on the picking schedule.

"It's amazing what we can do at this scale," he says, walking back to the SUV. "We manage a remarkable uniformity, which is a real skill and a testament to the technology."

The technology and scale are impressive, very different from the smaller wineries that pick their fruit by hand and often want individual blocks to express themselves. Ste. Michelle generally seeks a consistent style of wine that is approachable and accessible without aging. Most buyers will drink it immediately.

Getting back in the SUV, we follow the gravel road as it curves back down to the Canoe Ridge winery, with grapevines descending toward the Columbia. Back at the office, Brian points out a large whiteboard that displays what's happening in the blocks. It's a game plan for the harvest. It guides decisions about when to taste and when to pick. Block 24 will soon be added to that schedule.

Donning safety vests and glasses, we pass through the interior of the winery, a huge, cavernous building of gleaming stainless steel equipment, concrete floors, and high ceilings. We head outside where trucks bearing grapes from the vineyards dump them into a gargantuan stemmer/crusher. It's similar to our small, worm-drive stemmer/crusher at Les Copains but probably a hundred times larger, allowing it to process thousands of pounds of grapes. Canoe Ridge processes some twelve thousand tons—an astounding number—of grapes a year here.

The machine hums and whirs as it stems and crushes the grapes. The grapes then get transferred to a sorting table outside the winery that removes any leaves, stems, or other impurities. By this point, the grapes are clean, stemmed, and ready to be fermented—all with little in the way of human intervention and unlike the sorting tables I worked in at DeLille Winery and Betz Family Winery.

"Wine making is an attempt to control a natural process," Brian says, overseeing the operation. "We take the low-impact approach. It's all part of the grand strategy."

Obviously, there are places in the process where smaller wineries can minimize the impact further, but for such a large operation, it's amazing how well the fruit is treated and carefully controlled at each stage of the process.

We stop at one of the large fermenting tanks to sample the wine. Opening a small spigot at the base of the large steel tank, Brian pours two glasses. He and the other wine-making team members check the ferments regularly to see how they're progressing and when they need to be pressed.

"It's Cold Creek Cabernet," he says, swirling the glass and then sniffing it. "It's super fruity. I'm going to leave it on the skins. It's close, but I'm going to let it sit."

He dumps the wine into the drain. One more decision made.

We try a Syrah in a clay amphora, a modern update of an ancient wine-making vessel. They ferment the Syrah in whole clusters, including the stems, giving a distinct savory flavor to the wine. "Soy sauce comes to mind," he says, tasting the sample. "The perfect amount of stems varies by vintage, but too many stems distract from the flavor. This one is hitting on all cylinders."

I swirl and sniff. It reminds me of the wines from Vieux Télégraphe, a famous winery in the Provence region of southern France that I've visited several times. It, too, ferments with whole clusters, but its grapes are more savory while Washington's tend to be fruitier.

Finally, we visit the barrel room, the last stop in the wine's journey before it gets shipped. It's cool, dark, and quiet, and smells a little musty like an old cathedral. Barrels are stacked to the ceiling. The room can hold eighty-two thousand barrels or two million cases of wine, an ocean of juice.

"Now we put the wine in new barrels," Brian says, passing two workers pumping wine into a French oak barrel. "We use French oak and American oak." The type of oak and percentage of it forms a part of the recipe. French oak often imparts a smoky, toasty flavor to wine. American oak imparts a sweeter vanilla flavor.

"To conclude, there are lots of decisions to be made," he says, finishing up the tour. "When to pick? When to rack? How to combine it into a blend? What barrels to use? How can we figure out that puzzle?"

He gestures around the barrel room. "There's no one way to do it. Different decisions will lead to a different style. It's subjective. Different, not better."

Walking back to the office, he says, "I'm very proud of this place. I love to show people this place. People have some notion of wine-making principles, but it's different when it's blown up to this scale."

Thanking him, I walk out to my car, taking one last look at the immense facility and the vast estate surrounding it. It's a dazzling display of technology and human ingenuity on a very different scale from the smaller wineries I've visited. But it follows the same basic principles: no tricks, no additives, no manipulation. Just grapes into wine—but on a massive scale.

7

Picking Grapes

It is early fall in the Yakima Valley, with clear skies, temperatures in the high seventies, and clusters of grapes dangling tantalizingly from the vines. At Portteus Winery outside of Zillah, I'm armed with a twenty-gallon plastic bucket and eager to learn how to pick grapes—a critical part of wine making. I want to learn from the ground up and experience all aspects of the process, even unglamourous tasks such as picking.

I've corralled friends Tom Remmers, Scott Driscoll, and Bob Berg to help. Arriving at Portteus Winery, I shake hands with Paul Portteus, the tall, curly haired, easygoing owner. Today he's wearing a T-shirt, jeans, and running shoes.

"The market is down for grapes," he says. "This is the first year we may not sell all of our fruit."

This news is bad for growers like Paul but good for us. We get to pick these gorgeous grapes. Amateurs have difficulty getting the best fruit, which normally goes to bigger houses such as the ones I've recently visited.

Paul walks through the vineyard with us. The glaciated peak of Mount Adams rises above the farms and valleys and foothills to the south. His vineyard sits high on the slope of the Rattlesnake Hills. At 1,400 feet in elevation, Portteus is one of the highest vineyards in the state. His grapes are renowned for their richness and spiciness.

"It's all organic this year," he says. "We had to put some sulfur on the vines before the fruit set but nothing since then. There are some leafhoppers around, but they shouldn't be a problem because they only affect the vine if the grapes ripen late."

Dust rises from the road. It hasn't rained here since the spring, but Paul uses drip irrigation when needed. We walk past tractors, buckets, and plywood bins for picking.

"All of the fruit is hand-harvested, which means that it's handled more gently," he says. "Slow pickers can harvest two and a half bins a day. Good pickers can do three to five bins a day."

He calls out to one of his workers. "Which rows are you picking?"

The worker points to the row.

"Why don't you pick these two rows?" he points to two long rows. "Try to pick the vines clean."

"Sure," I say, eager to give this a shot. It shouldn't be that hard to pick a bin. How hard can it be?

He explains that the grapes grow on a bilateral cordon, with the trunk splitting into two canes that are tied to the cordon. The individual canes sprout off that, bringing forth leaves, vines, and bunches of grapes. Most of the clusters form about four feet off the ground, making it easy to pick them.

Paul hands me a set of clippers with blades resembling a bird's beak: long, narrow, sharp, and ideal for inserting between the vines to get the clusters. He uses his left hand to pull away the leaves and his right to clip the clusters and drop them into the bucket.

"They're about twenty-three Brix," he says. "That's good sugar content."

At first, we work slowly, figuring out how to do this. I learn it's easier to pick on one side of the vines and let the other guy work the opposite side. I pick one side; Scott picks the other. Tom and Bob pick the second row. We settle into a rhythm, hearing the snip-snip-snip of the clippers and the hollow sound of clusters hitting the bottom of the bucket.

The berries are small and sweet, with some browning, indicating it's time to harvest them. As we pick, we talk about the classes we teach at the university and how things are going with Scott's remodel. All is calm until a fight breaks out.

"Hey, you're filling up my bucket with leaves, Tom!" Bob says.

"It's the wind, man," Tom says.

"I'd be doing better without those leaves."

"Hey, is that your finger I just cut?" Tom asks.

"You'd know if it was my finger," Bob says.

Working opposite each other, we need to be careful. I want no fingers clipped today, though I nearly nicked my own left index finger with the shears.

The temperature rises. The sun beats down. Sweat beads on my forehead. We strip off our shirts. My fingers are flying, but the bucket fills slowly. When it reaches halfway, I drag it to the road, creating a dust cloud.

Leafhoppers and spiders are in the grapes but no black widows, I hope. Two years ago, we spotted a black widow just as it was heading into the stemmer/crusher. We had a special name for the vintage—Black Widow Blend.

Scott waits till his bucket is nearly full before dumping it into the bin. "Oh my God," he says, hoisting it to his shoulder. The bucket weighs about eighty pounds. This is true peasant work. Wine making may be full of poetry and romance, but much of it is grueling labor.

We keep going. Slowly, the buckets fill. A half hour goes by. An hour. I lose track of time. The silence is broken only by the sound of grapes hitting the bottom of the buckets.

Then Tom hits Bob with a grape.

Bob retaliates.

The peasants are revolting.

If you can't beat 'em, why not join 'em?

I toss a grape at Scott, who tosses one back at me.

Eventually, I call a truce. "We need to get our five hundred pounds."

Everyone goes back to work. Finally, we manage to fill a bin and get it hauled back to the weigh station. It tops out at 590 pounds; with 40 pounds subtracted for the weight of the container, it totals 550 pounds. Paul charges us for 500.

"Did we talk price?" he asks.

"Not yet."

"Well, another guy paid fifty cents a pound for them."

"That sounds fine."

"That will be two hundred fifty dollars. Oh, wait. You picked them. I'll deduct that. How does two twenty-five sound?"

Four guys working for two hours. That's about three dollars an hour. While Paul's other workers could pick two to five bins a day, we did only one—and that was with four of us picking.

I think I'll keep my day job. My back aches. My hands are sticky with grape juice. My legs are covered with dust. We transfer the grapes into plastic buckets and hoist them into the Eurovan for the crush.

8

Fermentation and Fruit Flies

It's fall and we're infested. Clouds of fruit flies fill the air. They buzz around my wife, Lisa, and me at dinner. They chew the chilies hanging in our kitchen, raining a fine red powder on the floor. They roost on the cupboards. They fornicate on the refrigerator. They threaten to overpopulate our house and our Seattle neighborhood as well.

"You have them, too," says Paula, a neighbor, swatting them away on our back porch. "They're bad this year!"

"Yes," I say, nodding, as if it's an act of God. But it's no accident that our house and neighborhood are infested. Some of Washington State's finest Cabernet grapes sit in our garage, fermenting into what I hope will be a stellar new vintage. Like most enophiles, I'd long fantasized about making my own wine—the romance, the poetry, the passion of it! Now that I've been doing it for a while, other words also come to mind—fruit flies!

My relationship with the insects is complicated. Their appearance coincides with the harvest, one of my favorite times of the year. Sometime around mid-October—I never know exactly when—I'll get a call: "The grapes are in." I have to drop what I'm doing, load plastic garbage cans into Tom's truck, and drive to the sunny eastern side of the state where the best vineyards—Sagemoor, Portteus, Ciel du Cheval—are located. Their names are beautiful to me. Arriving back in Seattle, I unload the buckets, wash the stemmer/crusher, and prepare for the crush.

As the members of Les Copains wine co-op arrive in the driveway of my house to begin crushing grapes, the first reconnaissance patrols of fruit flies appear. They buzz around me excitedly, making their distinctive right-angle flight patterns, intoxicated by the smell of the juice.

At first, the flies add a festive atmosphere to the crush. We dump container after container of grapes into the stemmer/crusher. We sample the sweet, sticky grape juice and previous vintages for comparison. Crush is a time to catch up with everyone and enjoy some wine together. By the end of the night, we're comparing our *vin du garage* to that of Château Lafite Rothschild. Our hubris knows no bounds.

After crushing the grapes into the large plastic tote, I add sulfite to the must to stun the wild yeast on the grapes. Some winemakers omit this step, allowing the wild yeast to ferment, but as amateurs we have a less controlled environment. We use cultured yeast for a more predictable fermentation.

Next, I dump packets of cultured yeast into a pitcher of grape juice and leave it overnight. The fruit flies hum crazily. The next morning, yeast foams over the top of the pitcher. Fermentation!

Taking the pitcher down to the garage, I dump it over the must, stirring up the fruit flies. The grapes are fat and shiny with a rich, sweet smell. Using a food-grade white plastic shovel, I mix the yeast into the grapes, getting sticky juice all over my hands. The temperature in the garage is around seventy degrees and is boosted by a small space heater. This helps the fermentation take off.

Later that evening, I return to the garage. The fruit flies circle excitedly, but the must seems not to have changed. I dip the shovel into the grapes, moving the heavy, sticky mass, hoping to energize the yeast. Have I done something wrong?

The next morning, fruit flies fill the garage.

"Hi, fellas!" I say, grabbing the shovel. The fruit flies buzz companionably, enchanted by the smell of yeast and fermenting wine. At this point, the fruit flies are a nuisance, not a danger to the wine. But as the fermentation slows, the odds increase that the *Acetobacter* bacteria, which the flies carry on their dirty little feet, could turn the must into acetic acid, or vinegar. That is a dire prospect since these grapes cost over a dollar a pound, and hours of labor went into crushing them. It still rankles that the bacteria once ruined a barrel of delicious 1995 Cabernet. To prevent this, I plan to press the wine as soon as the fermentation is finished.

The next morning, a scum of brownish bubbles appears on top of the grapes. The thermometer reads eighty degrees, which is a promising sign. The yeast is eating the sugar in the grapes, turning it to alcohol and carbon dioxide (CO_2). Churning the grapes with the shovel, I inhale the heady perfume of CO_2, yeast, and grape juice. Fruit flies circle around me, intoxicated by the smells.

Though I'm pleased with the progress of the fermentation, I know that problems can occur. If the fermentation doesn't take off, the wine can be ruined. Again, this is one of the reasons we used cultured yeast. Still, fermentations with cultured yeast can get "stuck," meaning the fermentation doesn't finish, making the wine prone to spoilage. I'd like to see more vigor in the fermentation.

The next day, the thermometer reads eighty-three degrees. The grape skins rise above the juice, creating a cap. Yes! The fermentation is picking up. I use the shovel to punch down the cap so that the skins have plenty of contact with the juice, ensuring the wine will have plenty of color and tannins.

Over the next few days, the thermometer's reading rises to the low nineties. The cap continues to rise. I hear a low seething sound as the yeast gobbles the sugar. The fruit flies can't contain their excitement. Clouds of them fill the garage, buzzing around me as I break the cap.

Things become strained. It would be fine if there were just a few of them, but an ever-larger cloud boils up to greet me. The parents become grandparents, great-grandparents, great-great-grand grandparents . . . It's hard to breathe without inhaling several generations.

The next day, the must reaches ninety-five degrees. I'm hoping it doesn't go any higher. If the fermentation is too hot, it can yield cooked flavors in the wine, making it taste burned or like caramel. Breaking the cap, I stir the liquid, hoping for the best. I turn off the heater to lower the temperature.

The following morning, the temperature falls to the low nineties. The cap begins to sink. As fermentation slows, the temperature drops. The fruit flies buzz ecstatically; they can't get enough of the grapes.

A week later, fermentation subsides, and we press the wine and transfer it into barrels. We submit the wine to rigorous "quality control analysis," meaning we taste it as we press it and speculate on its prospects, which improve the more we imbibe. The fruit flies dance around us as the dark purple liquid sloshes into the barrels. The spilled wine whips them into a frenzy. They land in it and paddle around, getting gloriously drunk on the last of the vintage.

The next morning, reality sets in. The grapes are gone. Chili powder appears on the floor. Clouds of flies swarm us at dinner, begrudging us every bite.

"There are a *lot* of fruit flies," Lisa says as they buzz around us. A biology major, Lisa likes fruit flies in moderation, but there's nothing moderate about the clouds circling around us. "In the genetics lab, we used traps to get rid of them."

"Good idea," I say, setting out traps made out of cups filled with juice and topped with plastic wrap with a hole punched in it. They quickly fill. These traps may have done the trick in the genetics lab, but they won't pass muster with the clouds generated by the grapes.

The buzzing reaches a crescendo. I can't even walk through the kitchen without parting a curtain of flies. Trying to open the fridge, I inhale several. That's it! I have had enough. It's time for the Shop-Vac, the nadir of my relationship with the fruit flies.

The cupboards and walls bristle with flies. As I pass the hose over them, they struggle to hang on but are quickly sucked into the maw of the machine. I am relentless, taking no prisoners as I make pass after pass over the flies. After several days of this operation, the buzzing stops. Peace reigns in the house.

"Thank God, they're gone," Lisa says. "I found some in the toothpaste."

I laugh nervously, hoping they really are gone.

A few days later, sipping our 1998 Cabernet-Merlot blend, which won a first place at the local county fair, I can afford to get philosophical about the flies. Sure, they're pests, but they're a small price to pay for having a home winery. I relax my vigilance, take my hand from over my glass, and look in to see several flies drowned in the liquid, reminding me that—like it or not—making wine and fruit flies are intertwined.

Next time, we'll do the fermentation at Tom's house.

9

Tom Cooks

In the aftermath of the Great Fruit Fly Infestation in our house, I sought to avoid divorce court by having Tom do the next fermentation at his house. With a detached garage and wide carport, his house made an ideal choice. The fruit flies could breed to their heart's content without infesting the inside of his house.

This arrangement pleased Tom. He enjoys wine making, too, and since I'm doing the cellaring and racking at my house, he wanted to do the fermentation and pressing at his place. Doing the fermentation allowed him to channel all the knowledge of organic chemistry and fluid dynamics he'd learned as a doctoral student. Though he works now in IT, he still likes to channel his inner chemist.

"The chemistry gives me a good understanding of what's going on," he says. "I understand what bonds are being broken and the CO_2 produced. It's easier to understand the process if you know chemistry.

"It's nice to have that control over it. It's my part of the process, and it's also my responsibility, and I take it seriously, monitoring and making sure it's doing okay. I want to do it the right way."

Wine making has evolved over the centuries. Once upon a time, you stomped the grapes, left them in a tank to ferment, and bottled them after. Now, we follow a detailed protocol.

We created our fermentation protocol by reading and consulting local winemakers. We added sulfite to stun the wild yeast, introduced cultured yeast, punched the cap, and managed the fermentation, adding water if necessary. This simplified process allowed us to make excellent amateur wine. But when compared with that of professionals, the product often came up short. It had great depth of flavor, but it lacked consistency and a silky finish.

We're amateurs, but we're always looking to improve. "I like that we're working on getting better," says Tom. "It's easy to get pretty good, hard to get really good."

To improve, I asked around and found an oenologist named Erica Orr, who came highly recommended. Erica graduated from UC Davis, the most famous enology school in the country, and apprenticed at wineries around the world, including Domaine Dujac, one of Burgundy's premier producers; Yering Station of Australia; and Napa's Cain Vineyard, Corison, and Rudd Estate.

Wine making is a mostly male-dominated field, but that's changing with talented young women like Erica. A consummate pro, she's slender and energetic, with big brown eyes that convey wonder at the mysteries of fermentation.

After we crush the grapes at Tom's house, he takes a fifty-milliliter sample of the juice to Erica's Woodinville wine lab. She analyzes the sample for its Brix, tartaric acid, pH, ammonia, and assimilable amino nitrogen levels—all key variables. She then comes up with a fermentation plan based on the analysis and our preferences. We want wine around 14 percent alcohol—not 15 or 16 percent—which sometimes means we must add water to lower the alcohol level.

She emails the results and formula to us. Tom now has his fermentation plan.

"It's reassuring to work with Erica's numbers because they've proven effective," he says. "I understand what the recipe means, and it's easy to follow. I like working with the directions because they allow you to interpret."

Erica recommends adding sulfite at the crush. It subdues wild yeast on the grapes. Some wineries ferment with these yeasts, but we prefer cultured yeasts for their ease of use and the flavors they produce.

After the grapes are crushed, Tom mixes sulfite into the must—without inhaling it.

"You don't want to burn your lungs with the concentrated meta-bisulfate," he says. "The only solution I've found is to fill up your lungs like you'll be underwater, then hold your breath. It can be hard to do, because usually you're pretty drunk at the end of crush."

After adding the sulfite, Tom adds a gallon or two of water to hit our target of 24.5 on the Brix scale. On day 2, he introduces the cultured yeast and GO-FERM, a yeast hydration nutrient. Erica suggested this step, and it's made a big difference in quality.

Day 3, Tom adds more yeast nutrients to keep things rocking.

"The joys of yeast nutrient!" he says. "Ferment takes off very quickly, so you need to account for that in adjusting the timing. For the 2019 ferment, I only put a tarp over the tote and used no insulation of any kind."

The grapes ferment in a tote, or a large, insulated plastic bin. As the fermentation progresses, the yeast devours the sugars in the grapes and gives off heat and CO_2, which can be toxic in confined spaces.

"You don't want to kill yourself with CO_2," he says. "Once we fermented in a tightly sealed garage, and I went inside too quickly. I felt dizzy for at least ten minutes. CO_2 kills!"

Now, when punching the cap, he removes the tarp covering the tote, allowing the CO_2 to escape, and goes inside the house for a cup of coffee. Then he returns to do the punch down.

As the fermentation takes off, the grape skins rise to the top of the must, forming a cap. They need to be punched down at least twice a day. This mixes the grape skins into the must, dissipates bacteria and mold, and adds color, flavor, and tannins.

"Punching the cap takes a lot of force," Tom says. "My technique is to go along the sides and push the plastic shovel along the edges of the tote—flat side against the inside wall, angled at about fifteen degrees. You get great leverage, and a lot of CO_2 is released. Then you have to punch the middle, but it's much easier once the edges are done."

As with so much in wine making, he takes pleasure in a job well done. "It's always satisfying after a good cap punch, seeing all the grape skins nice and wet."

As the grapes ferment, the temperature rises, usually getting up to a peak of eighty-five degrees or so. This allows full extraction of the color, tannins, and other components of the grapes, making the wine rich and robust. If the ferment gets too hot, it can lead to cooked flavors; if it doesn't get hot enough, the fermentation can get stuck, leaving a lot of residual sugar, which we don't want as we're making dry wine.

"The temperature is tricky in home wine making, especially if the ferment occurs to open air in the carport," Tom says. "The fermentation process produces heat, but colder air temperatures can slow down fermentation."

A handyman with a gift for improvisation, Tom builds a tent around the tote during cooler years and uses a heater to keep fermentation going. "I enjoy the challenge of keeping the temperature consistent," he says.

As the ferment slows, he keeps an eye on the temperature and specific gravity of the must. He measures the latter with a hydrometer, which tells him how much sugar remains.

"An accurate hydrometer is useful when you get close to being done, but looking at how much the cap is pushed up is the best indicator," Tom says.

I check in with him during the ferment, but I leave him to get it done. It's his baby, and I trust him to do it right. We may disagree on a few things, such as how thoroughly to wash the carboys for the pressing, but we work well together as a team.

"I think we have the same vision," he says. "We both want good wine. We're the type of people who enjoy hiking for eight hours, so we're willing to put in the hard work in return for something good. I would like more cleanliness, but I'm not staging a revolt or breaking off into my own sect. We've been making wine together for twenty-five years. How many other relationships have lasted that long?"

I'm grateful for his commitment. It's a team effort, and he's a critical part of it. When I get the word from Tom, I send out an email to the Les Copains group, inviting them to the pressing, which is the next key step in the grapes' journey into wine. He's always relieved that his part of the process is complete.

"It's a tag team," he says. "I've done my part, and then I hand it off to you. There's satisfaction of a job well done."

The pressing is always exciting as we get to taste the year's vintage for the first time. Every year we buy grapes from the same vineyards and treat them the same way, and yet there is always variation. What will this year's juice be like?

10

Pressing

The Merlot Mobile

Tom's house sits in a leafy ravine north of downtown Seattle and is surrounded by rhododendron bushes as big as trees. After backing my Subaru into his driveway, I get out and walk over to the plastic tote to look inside. The cap has fallen, and the surface is flat. Flecks of purple foam cling to the sides, indicating fermentation is nearly finished. Tonight we plan to press the wine off the skins and transfer it to the thirty-five-gallon glass carboys scattered around his carport.

This moment is always exciting. The yeast has transformed the grape juice into wine. The alchemy is complete, though the wine is in its infancy. I dip a cup into the must and taste it. It tastes chalky, yeasty, and fizzy; the flavors are there, but it will take time to bring them out.

Tom scurries around, setting things up. He takes out the Italian-style ratchet press, a classic piece of machinery used by generations of winemakers from garagistes like us to small-scale commercial operations. He cleans and sanitizes the curved wooden slats and the metallic base that will come into contact with the wine.

I wash out the carboys twice.

"Do it once more," Tom says. This is part of the ritual: I think they're clean; Tom wants them cleaner. I give the carboys another rinse.

Soon our Les Copains members arrive. We have a crack team helping us tonight. First is Andrew Rice, a tall, carrot-topped anesthesiologist, who is meticulous with the fermentation airlocks, making sure that the wine doesn't get oxidized or exposed to air.

Gregory Heller, a diehard foodie, serves as our utility infielder, helping with filling carboys, pressing skins, cranking the press, and hauling away the pomace, a cake of dried grape skins that appears after we've pressed the grapes.

Bruce McLachlin, Tom's neighbor and fixer of the stemmer/crusher, brings over a hot plate of pizza he cooked on his Big Green Egg grill.

Theresa Jurotich, our label maven, appears with a bottle of wine and a plate of cheese.

Another part of the ritual is to bring wine and food to sample, including older bottles of Les Copains to see how it ages. Everyone brings something, whether bread, wine, or cheese, creating an impromptu holy trinity of a meal that underlines the sacramental quality of the occasion.

Tom trundles his grill out to the driveway and fires it up. He lays on some sausages and steaks. The party is started!

Everyone dives into the food. We have the inner circle of Les Copains with us tonight, all of us experienced, fun loving, committed to making wine. Others join us when they can, sometimes asking if we can change the date to accommodate their schedule. No! This is wine making, dammit! The grapes dictate when they're ready to be pressed.

"I enjoy the camaraderie of the people," says Tom. "People learn that wine making is not a big snotty thing. It's simple and elegant and yet complex."

After we toast the harvest and eat, we get to work. We've got a lot to do tonight. Tom puts the two halves of the wooden basket around the base of the press and locks them in place with steel pins. He uses a pot to scoop the must from the tote and into the press. The juice gushes through the basket and into a bucket underneath that catches free-run juice, dark, purple, delicious.

Once the basket is nearly full of grapes, Tom places two half-moon-shaped wooden caps on top and adds crossed two-by-fours for leverage. Then he threads the ratchet down the vertical shaft until it rests on top of the two-by-fours. He starts working the ratchet, which pushes the two-by-fours and wooden caps onto the grapes, gradually extracting the juice. This is the pressed juice, which is slightly more acidic than the free-run juice.

The juice sloshes into the bucket below. Andrew replaces the bucket with another and then takes the first bucket to fill a carboy, using a funnel. Leaving an inch of air space at the top, he then fills an airlock three-quarters of the way up with water and attaches it to the top of the carboy. This will ensure that little air gets in.

After putting a blue tarp into the back of the Subaru to avoid spillage, I pick up the first carboy, using gripper gloves to avoid dropping it. The slippery carboy weighs over forty pounds, making it a grunt to get it into the Subaru. The carboy bubbles as I set it in place. I load the other carboys into the vehicle, stacking them right next to each other while trying to minimize any movement. I don't want even one of them to break because of the loss of the liquid and the difficulty of cleaning the car. Eventually I cram fifteen into the car, or half of tonight's production, weighing over six hundred pounds. The Subaru becomes a very low rider.

While Tom and the others continue pressing, I drive back to my place, going slowly, carefully, not wanting the carboys to break. They shake and rattle in the back as I take a right and head down Fifteenth Avenue. Then I creep over the Ballard bridge, along Nickerson, and then up Queen Anne hill. I go slowly and back up into my driveway. The Subaru scrapes on the sidewalk, revs high, and gradually eases up the driveway.

I carry the carboys into the garage and place them next to the wine barrels. I put them in a row, getting them ready for the racking. I'll leave the carboys for a week or so on the cement floor, allowing the fermentation to finish and their yeast and sediment to release to the bottom.

Then I head back to the pressing. When I return, the crew is still working on the pressed juice. Everyone is taking a turn at the press.

Eventually we finish. After carrying the remaining carboys into the back of the Subaru, I'm worn out; no need to lift weights tomorrow. I let Tom and the others wash out the fermenter and press as I drive back home with the carboys, trying to keep them from breaking. Again I drive slowly, conscious of not wanting to make them jostle in the back. I reach the base of Queen Anne hill. It should be smooth sailing.

But as the car hits a bump, the carboys clash together, and I hear the sickening sound of shattering glass.

"Shit!" I can imagine the mess in the back. I'm close to home, so I keep driving. By the time I arrive at my house, the smell of Merlot is overwhelming. I park and take a look in the back. Sure enough, one of the carboys has shattered, sending five gallons of wine washing over the back of the car. The blue tarp caught some of it, but the rest floods the back of the vehicle. I clean up what I can, pick up pieces of the broken glass, and bring the rest of the carboys into the garage.

Though I have cleaned the car many times since, I can't get rid of the smell. Ever since that incident, especially on a warm day, a fragrance wafts up from the rear of the car, smelling like—you guessed it—Merlot! I've christened the car the Merlot Mobile.

And when someone asks, "What's that smell in your wine? I can't quite place it."

I say, "It smells just like Subaru."

PART 3

Oregon

11

Learning the Secrets of Making Killer Pinot Noir

"Terrific tannins!"

"Fantastic fruit!"

"Bodacious berries!"

"Chewy chocolate notes!"

The wine jargon is flying fast and loose at the International Pinot Noir Celebration in McMinnville, Oregon, where many of the finest Pinot Noir producers from around the world and a crowd of Pinot Noir aficionados like me have gathered to taste the best of the recent vintages. This highbrow bacchanalia takes place at Linfield College in the heart of the Willamette Valley, where thanks to a soil and microclimate similar to that of Burgundy, France, some of the best Pinot Noir grapes in the world are grown.

Nicknamed the "Big Slurp" by industry insiders, the festival features three days of tastings, tours, seminars, and amazing meals. In addition to this three-day gastronomic and enological extravaganza is a Sunday afternoon affair, which is easier both on the waistline and the wallet, and still allows you to taste all the wines, converse with the vintners, and search for suitable descriptions of their remarkable offerings.

My wife, Lisa, and I attend the Sunday event. As with most of the others, we've just enjoyed a sumptuous meal of spit-roasted baby goat, grilled chicken breast, grilled sausages, grilled vegetables, prosciutto, terrine, grape leaves stuffed with chèvre, and a decadent assortment of truffles, eclairs, and seasonal tarts. Keep in mind that this is the light, afternoon menu, which barely avoids gluttony.

Having finished this extraordinary meal and sampled several fine Pinot Noirs along the way, I'm now ready to get serious about tasting. I begin circling the hall with my official International Pinot Noir Celebration Schott Cristal goblet and spiral notebook, searching for that perfect bottle of Pinot Noir. Since the wine has so many different styles, this quest is somewhat absurd, like looking for the last unicorn, but it's certainly a pleasant one.

Unlike Bordeaux and many other famous red wines, Pinot Noirs often come from single vineyards, and since the grape is notoriously temperamental, everything has to be right to create an outstanding wine. To produce an extraordinary Pinot Noir year in and year out is very difficult. Soil, climate, weather, vinification—and the position of the planets, it sometimes seems—all have to align to make memorable wines.

When a winemaker succeeds in coaxing the best out of this notoriously fussy grape, the results can be remarkable. The finest Pinot Noirs are rich, perfumed, infinitely subtle wines that grow in taste and complexity with age. If money is no object, you can simply buy your way to enological perfection, shelling out big bucks to purchase a 1985 Domain de la Romanée-Conti or Gevrey-Chambertin. But if you can't afford such extravagance and do enjoy the process of discovery, then you can attend tastings such as this one and find wines that can be just as magical and much more affordable.

I began my fascination with Pinot Noir while in graduate school, at a time when I had little if any extra money and certainly not enough to afford a fine Burgundian Pinot Noir. I bought a sale-priced, eight-dollar 1985 Pinot Noir from California's Gundlach Bundschu Winery and found it a revelation. I surprised myself by buying a case of it. I'd never purchased a case of wine before and felt ridiculously indulgent in doing so. But the wine was so wonderful and the price so low, how could I resist?

On trips to Europe over the years, I've made pilgrimages to Burgundy to sample some of the world's best Pinot Noirs. I've visited the Marché aux Vins in Beaune, France, several times, where for ten dollars you can sample twenty to thirty different Pinots, including some of the grand and premier crus. I remember in particular a 1978 Chambolle-Musigny that sang the most exquisite melody in my mouth. I was tempted to put the entire bottle to my lips and drink straight from it without the needless interference of the tastevin. I barely restrained myself.

Every connoisseur of Pinot Noir has had these conversion experiences, small epiphanies on the path to perfection. While the quest for the perfect bottle of Pinot Noir may be impossible, experiences such as these make you believe it may be possible and prod you to keep searching to discover it. It's just such an experience that I'm searching for today.

"Concentrated fruit!" I hear someone yell from the other side of the room, and I immediately head over to see what all the commotion is about. A number of people are standing around the table where Christophe Perrot-Minot is pouring the Charmes-Chambertin Grand Cru. I proffer my glass, he pours me a bit, and I take a whiff of the dazzling bouquet. I wash the wonderful garnet-colored liquid around in my mouth. It verges on perfection.

"Great structure and balance!" Someone shouts from the left. I savor the Charmes-Chambertin for a few more seconds and move over to the next table. There, Geoff Bull, the big, bluff, hearty representative of Freycinet Vineyard in Tasmania, pours me his latest Pinot Noir with one hand and gives a demonstration of how to fend off a Tasmania devil with the other.

"You have to watch out for them," he advises me. "They'll eat your shoes."

I thank him for the self-defense advice and for his fine, well-balanced wine. His Pinot is much lighter than the Charmes-Chambertin, but it nears perfection in its own way.

Up to this point I've mainly tasted French and other foreign Pinots. This strategy is deliberate, because buying these wines locally is difficult and because I prefer a sharper, more acidic wine. I find many of the Oregon Pinot Noirs, which are lush, rich, open, new-world wines, sometimes lack a bracing edge of acidity to balance their youthful innocence and enthusiasm. At least one exception to this gross generalization is Domaine Drouhin, a 135-acre Oregon winery owned by the Drouhin family, one of the premier producers and sellers of French Burgundy. I make my way over to see if their latest wine is as good as previous samples I've tasted.

I have to wait in line because Bill Hatcher of Domaine Drouhin (now with Rex Hill) is mobbed with wine drinkers. He pours with both hands to keep up with the demand. Visitor after visitor stops by to taste his winery's elegant Pinot Noir. When I finally get a sample of the latest vintage, I'm not disappointed. It has a fine balance of acidity and fruitiness. The wine is as austere and formal as the finest French Burgundies yet as fruity and flavorful as the best Oregon Pinot Noirs. It's a wonderful marriage of the old and new worlds.

"Wonderful bouquet!" someone effuses from behind me. I head over to taste other Oregon Pinot Noirs. I sample a fine Amity Willamette Valley Winemaker's Reserve; a rich, complex Argyle Willamette Valley Reserve; and a dense, smooth Knudsen Erath Willamette Valley Reserve.

As the wine flows and the din increases, my taste buds dull, and everything begins to blur. Originally, I wanted to sample all the wines, but I decide that even if it's possible, it's not desirable. No, instead, I'd rather relish the amazing food and wines I've already tasted, contemplate how close some of them came to perfection, and save further tasting for another day.

The great thing about seeking perfection in Pinot Noir is that you never quite attain it, allowing the search to continue. I'll probably never find the perfect bottle of Pinot Noir, but that allows me to continue my quest, tasting yet another bottle and occasionally experiencing the epiphanies that bring me back to the Willamette Valley again and again.

I have great respect for Pinot Noir producers and their quest for the perfect Pinot. Early in my wine-making career, I managed to get my hands on some Oregon Pinot Noir grapes—not an easy task as they always seem in short supply. Tom and I stemmed and crushed the grapes, fermented them, and then aged them in carboys. The wine was light and fruity with a hint of cranberry and a subtle tang of soap. Palmolive?

It never matched the lushness, subtlety, and complexity of the great Oregonian or Burgundian Pinots or, for that matter, our Cabernet and Syrah. The difference could have been the fruit or our ignorance about how to treat the grapes. Pinot Noir is notoriously finicky; if you can turn it into a stellar wine, then you have the chops to make any wine.

I decided to make another trip to the Willamette Valley, not just to taste wine, but also to learn how to make it. I managed to join the volunteer lists of three master Pinot producers: Mike and Mikey Etzel, two strong-willed characters who run Beaux Frères Winery, one of the best in Oregon; Tony Soter, the owner of Soter Vineyards, a standout producer of Pinot and sparkling wine; and Ken Wright, an Oregon Pinot Noir evangelist and the guiding spirit behind creating the six AVAs in the Willamette Valley.

All four helped pioneer Oregon Pinot Noir while standing on the shoulders of earlier figures such as David Lett of Eyrie Vineyards. In 1979 Lett sent a sample of his South Block Reserve Pinot Noir to the Gault-Millau French

Wine Olympiad. Lett's 1975 Pinot Noir stunned the French establishment by winning first place in the competition, putting Oregon on the world wine map.

Drawn by the chance to cultivate this exalted grape, winemakers flocked to the region, including David Adelsheim, Dick Shea, Dick Erath, Doug Tunnell, Patricia Green, Steve Doerner, and many others. They paved the way for the current industry, with over one thousand wineries calling Oregon home.

12

Beaux Frères

Avoiding Getting Lee-ed

Beaux Frères winemaker Mikey Etzel scoops a gob of purple goo from the bottom of the wine press. The goo contains fermented grape skins and lees, or dead yeast, the by-product of the fermentation.

"Is this reductive?" he asks his assistant Omar Perea, shoving it toward his face so he can smell it.

"Yes," he says.

Mikey puts the grapes in front of me. "Is it reductive?"

I sniff the goo, which smells strongly of yeast and other scents I can't pinpoint. I've never smelled the lees of Beaux Frères wines, though I've smelled plenty of Les Copains' lees. It smells normal to me for this stage of fermentation.

"Yes," I say, tentatively.

"Are you sure?" says the lanky, sardonic winemaker who wears a pile jacket, wine-stained pants, and several days–old stubble. To give me a better sniff, he feigns shoving the goo in my face. I've been caked before but never leed, if that's even a word.

"I don't know you well enough to do that," he says, grinning and tossing the goo back in the wine press.

"That will come later," I quip.

Welcome to the craziness of the crush, when long hours, backbreaking work, and near exhaustion make winemakers giddy and loopy. Tracking reduction, or the smell of fermenting yeast, is one of the many tasks performed during crush, when all the year's labor comes to fruition in a short window. I'm excited to participate in this process at Beaux Frères, one of the most renowned wineries in Oregon. Founded in 1986 by brothers-in-law (*beaux frères* in French) Michael G. "Mike" Etzel and wine critic Robert

M. Parker Jr., the property consistently earns top scores from wine critics around the globe.

Located on an eighty-eight-acre former pig farm in the Willamette Valley south of Portland, Beaux Frères produces world-class Pinot Noir from organic and biodynamically grown estate grapes. Etzel originally planned only to grow grapes, not make wine. But finances dictated otherwise, so he dived into learning the craft. He proved a fast learner: His first Beaux Frères release in 1991 earned some of the highest marks for Oregon wine and fetched an impressively high price (thirty-four dollars). Etzel was on his way.

Since then the winery has continued to produce stellar wines, earning high scores from industry publications such the *Wine Spectator* and increasing its production to some ten thousand cases per year. Such excellence doesn't happen by accident. The wiry, energetic Etzel exudes competence as he works diligently to plant vines, care for them, harvest the fruit, and craft wines of "strong character," as he likes to say.

In the early years, the wines showed the influence of Robert Parker, the American wine critic and part owner. He encouraged Mike to make wines in his preferred style: big, powerful, highly extracted.

"My brother-in-law, Robert Parker, wanted low yields and aging in new François Frères barrels," Mike says. "The wine consumer was looking for power and structure and fruit. Gradually, our wine evolved into more nuances and ethereal aromatics and less alcohol."

Despite the evolving style, the wines continued to earn stratospheric scores, largely due to Mike's exacting approach. This perfectionism served him well in the vineyard and winery, but created clashes when he sought to bring his family into the business. Born in 1986, son Mikey showed particular promise. He grew up on the family farm, where he learned how to tend the vines. He left to study at Oregon State University's enological program, graduating in 2008.

Following graduation, Mike joined nearby Willakenzie Estate as the assistant vineyard manager. Under the tutelage of Daniel Fey, who ran the 130-acre vineyard, Mikey sought to achieve a similar level of quality at Beaux Frères.

"After that I wanted to go to a smaller estate," he says. "It was a grind in the vineyard. I wanted more comprehensiveness."

The following year, Mikey and his brothers Jared and Nathan founded a small-batch production project called Coattails. The name served to acknowledge the many individuals who had helped them on their journey as winemakers. In 2011 Mikey moved to the Ribbon Ridge AVA winery Brick House Vineyards, where he managed a twenty-acre biodynamic vineyard. This experience convinced him that it was the best approach to farming.

In 2013 he became winemaker for a start-up operation called Chapter 24 Vineyards. The project was guided by a consulting winemaker Henri Jayer of Vosne-Romanée, from whom Mikey learned an enormous amount. The position proved to be a new challenge, putting him in charge of a twelve-thousand-case production that sourced grapes from across the entire Willamette Valley.

After this extensive experience, Mikey returned to Beaux Frères as head winemaker and viticulturalist in the 2016. His time away from Oregon gave him a vital global perspective on the wine business.

"We are competing not just in Oregon," Mikey said. "We are a world organization competing on the world stage. Let's not be closed-minded in our approach to wine making and marketing."

Today Mike serves as president and founding partner, overseeing the whole company: sales, marketing, and budgeting, as well as picking the fruit. Mikey is now the chief executive officer. The collaboration between father and son has paid off handsomely at Beaux Frères, with stratospheric wine scores and an enviable customer base. Though their palates differ, Mike and Mikey combine forces to craft stunning wines.

"It's a higher sense of meaning to work in a family business," Mikey said. "There's a blurring of lines between the personal and the professional. It's way more fun and more meaningful."

But family businesses can be challenging. The 2017 harvest proved a stress fest at Beaux Frères in Newberg, Oregon. The winery's yields, or harvestable grapes, were 30 percent higher than normal, requiring that they process twelve thousand cases of wine in a facility designed for ten thousand. All this came on top of twenty-hour days and backbreaking work.

Mike and Mikey argued about the best way to determine yields. Mike favored an impressionistic approach, walking the rows and estimating yields.

Mikey favored a more methodical way of judging yields. The long hours and exhausting work caused the argument to boil over.

"I quit," said Mikey finally. "I'm out of here." He walked back to his house, intending to leave the winery and move on.

After collecting himself, Mike swallowed his pride and walked over to his son's house and apologized. "Harvest can be painfully stressful," said Mike. "With my crew, if I say, 'Jump,' they say, 'How high?' Mikey would question why. In the early years, Beaux Frères was basically me. Over the years I've learned the art of delegation."

After the 2017 harvest, they decided a division of labor would work best.

"After I walked out, we came to a better consensus," Mikey said. "During harvest, we'll collaborate on the first pick. Once we get into the big pushes and strategy of the picking, my dad is going to be in the vineyard, managing the picking. I'm going to stay in the cellar. I'm in charge of processing. In 2018 we implemented this plan, and it minimized the stress."

Despite the challenges, the Etzels see clear reasons for running a family winery.

"There are many advantages," Mike said. "We own a percentage [of Beaux Frères], but we are owned by the French [Maisons & Domaines Henriot]. One reason they purchased us is because we are a family business."

The Newberg winery is located in a converted green barn at the base of Ribbon Ridge amid a grove of oak trees. It fits neatly within the rolling farm country of the Willamette Valley and houses in back the bulk of the winery: fermenting tanks, tasting room, offices, and barrel storage. A large roof projects out over the front where the stemming and crushing takes place.

I join Mikey and Omar as they monitor the juice coming out of the membrane press. The press ensures gentle treatment of the fruit. They've programmed it to squeeze the grapes over two hours, releasing the juice gradually to avoid excessive tannins.

When the grapes are pressed, the free-run juice gushes out first, falling into a stainless steel container. Then the grapes are run through the membrane press to gently squeeze the remaining juice. The trick is to do this without adding excessive tannins. The skins and seeds contain tannins, some of which are appealing, giving balance to the wine; but too much tannin makes the wine astringent and mouth puckering. There's a fine line between balanced and overly tannic. Think Beethoven's "Moonlight Sonata" versus AC/DC's "Highway to Hell."

Mike stops in to check on the pressed wine. Wearing jeans, a denim shirt, and work boots, he grabs a glass and sniffs the wine. Though his primary responsibility is bringing in the fruit, he tastes throughout the harvest, checking to make sure it's on track.

Having seen many harvests, Mike remains anxious until he makes the decision to pick. "Then it's full bore ahead," he says. "Once it begins, things tend to smooth out. You've got all your vats prepared, harvest crews and interns ready to go. You're anxious to get into battle."

"How is the harvest going?" I ask.

"We started out with a late frost and thought we were in deep trouble," he says in a gravelly voice. "The frost killed a lot of primary buds but not the secondary buds. I thought we were in trouble, but the plants seemed to do well."

After that, the weather turned warm and dry, with cold nights and dry days. It rained once during this time. "That rain was God sent because it helped the plants ripen with flavor," he says. "Then cold nights and moderately dry and sunny days enabled us to harvest when we wanted to."

He pours a splash into a glass, sniffs it, and nods approvingly.

"I like reduction myself," he says, swirling the cloudy new wine around in the glass. "It means that the yeasts are consuming the sugar. We don't feed our yeast with nutrients. We like them to strain and struggle to build complexity. Reduction more frequently occurs in old-world wines. The winemakers of today are taught that's a flaw. I think it's a beauty mark. But I'm a weirdo."

He hands me the glass, and I sniff the wine, trying to get a sense of what he means. "How would you describe the smell?" I ask.

"You're looking for things aren't there," he says. "The wine's in a very confused state. It's going from grape juice to fresh wine with a lot of yeast and turbulence. If it smells like airplane glue or hair dye, that's a problem," he says. "I smell reduction and it's good reduction. Some old-timers say old Burgundy tastes like shit. I think it smells like a beautiful woman's armpit after a week of not showering."

I take the glass back and sniff again. It smells strongly of yeast, but I don't notice any off flavors, whether airplane glue, excrement, or eau de armpit.

"If the wine smells good, it almost always tastes good," he says. "The wine will not surprise you. Always smell the ferments. Get a glass, smell and taste, and you can see the evolution. First free run. Then taste the pressed juice as it goes through its cycle, looking for tannins, color, and smell."

I inhale again, detecting some of the fruit behind the yeast.

"I want complexity in the aromatics," he says. "If it has no smell, you've gone too far in the pressing. You have a valve on the drip pan. Know when you're getting toward the end. Close the valve on the press pan. After a few cycles, you get a feel for the vintage. As the wines settles a month later, you smell and taste it, you begin to understand the vintage."

I've never gotten so detailed an explanation of pressing wine. At Les Copains we've always just pressed till the liquid stops flowing. He's provided a master class in analyzing newly fermented fruit.

Mikey stops by to taste the wine. "It's pretty stinky and skunky," he says taking another sniff of the wine. "But the reduction is not too bad. It just needs some air."

"I like the reduction," Mike says. "It's good reduction."

Mikey nods and leaves to operate the forklift. They seem to appreciate each other's opinion even if they don't always agree.

"Everyone tastes and smells different," Mike says. "I trust his palate and judgment on wine. Then you measure degrees. How do you draw the line between good and bad reduction? I may be more tolerant of reduction because I drink more old-world wine."

Mike excuses himself to get back to bringing in the fruit. Mikey returns to taste.

"It's a good exercise," Mikey says of his and his dad's kibitzing over reduction. "We hash it out in a way that's more natural. There are a lot of confounding variables. He's crazy, but I'm crazy too."

Mikey jumps on the forklift to bring in another tote of fruit. I walk over to take a look. It's nearly immaculate with few leaves, sticks, or anything else. A crew of four diligently sorts it though they don't seem to have a lot of work to do.

"The fruit is pretty clean," he says, picking through the clusters and grabbing one to taste. "It's beautiful fruit with good acids. We've captured freshness in all these wines. It's what we work for all year."

I taste the grapes along with him. I notice some of the freshness he's talking about, more red fruit like raspberries and cherries.

"When I came on board, I changed a few things," Mikey says. "I changed to early picking and a more gentle extraction, pumping over instead of punching down on the cap. Wines of my tenure have more primary fruit and are a little softer."

Taking a break from the press, he shows me around the back of the winery, where tanks are fermenting and the smell of yeast fills the air.

"I brought changes I'd learned from Mark Tarlov and Burgundian consultant Louis-Michel Liger-Belair at Chapter 24," he says. "We have a warmer climate in Oregon than in Burgundy. We need more pump overs and less extraction. We're still a young industry though. We haven't been doing this for hundreds of years."

It's been seventy-five minutes. We head back outside to check on the press. It continues to gently squeeze the grapes, releasing a trickle of juice. We smell and taste it. The wine is becoming less reductive. If pressed too long, bitter tannins can come out, released from the seeds. But I don't taste any of those yet.

In between tasting, Mikey mentions that three weeks ago, Artémis Domaines, an investment group, bought Maisons & Domaines Henriot. Beaux Frères has changed hands yet again.

"I'm excited about it," he says. "It's a strong brand, and I'm excited to be here. It's a great new chapter. We can focus on estate wines, lower production, and making the best Ribbon Ridge Estate AVA. It's another feather in our cap."

Though deeply rooted in the Willamette Valley, the winery also has clearly acquired an enviable international reputation.

"What's the secret to your success?" I ask.

"A great site," he says. "And strong effort in the vineyard. We take great pride in what we're doing. We're vignerons, not winemakers. The cellar is a midwife. We gently coax the juice from the grapes. We just give nature a little helping hand. Growing grapes is the hard part."

He jumps on the forklift to bring in another tote of grapes and then returns. Some would suffer from task overload, but he seems to handle it fine.

"It's a collection of decisions that makes the wine," he says, continuing the conversation. "There's no recipe. No secret. We don't compromise on what we feel is right. This is our life. We eat, breath, shit, and bleed this place."

Turning to Omar, he says, "We have a great team, a great company culture. We're not corporate. We're a family culture. We have no big office in Portland. Everyone is a part of what we do."

He sees the farm and the people as an organism. The property is not just a winery or a vineyard but also a whole ecosystem, which includes the folks who work here. This sense of sustainability includes the vineyard, the winery, and the whole area around it, making sure it can be passed to the next generation.

"What did you learn from your dad?" I ask.

"Work ethic," he says simply. "Use feelings to make decisions."

I think immediately to tasting their 2018 Beaux Frères Vineyard Pinot Noir, which earned a ninety-five rating from the *Wine Spectator*.

"It has a telltale aromatic nose," Mike said of it. "There's no recipe. It's a dynamic thing. We make decisions based on feelings and intuition."

Mikey obviously has the schooling to make decisions based on the numbers, and I'm sure he consults them when making critical decisions. But as he's shown today in tracking reduction, it all comes down to taste and an intuition of what the finished wine will be. This sense is something I'm still struggling to master.

"It's almost there," he says, taking another sip of the pressed wine.

"How is the harvest going?" I ask. It's been warm and dry over the last few days, but rain is due shortly. Will Beaux Frères be finished by then?

"It's tranquillo," he says, noting that the last of the fruit is picked.

One hundred twenty minutes. Omar sniffs and sips of the juice. "Time to stop."

"All right!" Mikey says.

"How does it smell?" I ask.

"Smell for yourself," he says.

I grab the wine glass and take a sniff. It smells of yeast but also of tobacco. My sense of smell has sharpened. I've learned a lot during my short time here (and managed to do it without getting leed!). I understand the need to smell the wine throughout pressing so that I will know when to stop. Smelling and tasting the new wine and tracking it through its various stages have proven invaluable. I also now see how this hands-on approach is key to the success of the winery as every vintage is different.

Omar scrapes the last of the pomace out of the press, his hands purple with lees.

"We're all done," Mikey says with a hint of relief in his voice. He washes out the membrane press and gets ready for lunch. "The fruits all in. The fat lady has sung."

13

Tony Soter

Pinot Noir as an Expression of Place

Storms line up off the Oregon Coast, waiting to march inland to drench the Willamette Valley. The gun-metal gray clouds gather on the horizon, moving toward the wine country.

Fall rains can decimate a harvest. The grapes take on water, diluting them and upsetting the sugar-acid balance, especially if the soil is poorly drained. The berries swell and split, causing mold, mildew, and spoilage, limiting or even ruining the crop.

Local wineries scramble frantically to get the fruit picked before rains arrive. But at Soter Vineyard's Mineral Springs Ranch, south of Portland, the mood is calm and purposeful. With over forty harvests under his belt, owner Tony Soter has a plan for getting the fruit picked before the rains fall. He's the calm at the center of the storm.

"This is the peak day of the season," he says, showing me around the Mineral Springs Ranch vineyard. "We're nearing the end. We need to finish."

Workers bustle around the winery with unhurried haste; they seem to know what they're doing and how to do it. They're closing in on one of the weirdest and most challenging harvests ever. The coldest April temperatures in eighty years seemed to doom the crop. A warm and dry July and August kept the vines on track. Fall rains have a history of spoiling the harvest, but so far they have held off.

"It's just part of farming," says Tony philosophically, who experienced his first harvest in 1975. "Every week Mother Nature deals you another set of cards in the poker game of farming. It's never boring. Never the same. You learn to adapt. The wines are different every year, with a character stamp of the vintage."

Tony is a compact, stocky man with a calm, deliberate manner and an easy gait. He wears a tan cotton shirt and pants, a green baseball cap emblazoned with a wine glass, and comfortable boots suitable for walking the vineyards.

I accompany him as he makes his rounds. I love the energy and excitement of the harvest and look forward to seeing how Tony orchestrates it.

We walk a gravel road toward the winery, which crowns a hill outside of Carlton and includes a biodynamic farm, vineyard, and tasting room framed by tall Douglas fir and oak trees. Soter's wines, food, and hospitality attempt to capture the essential character of the place—an elusive but intriguing goal, as I'll see during my visit.

Tony introduces me to Chris Fladwood, the intense, bearded winemaker who tends the fermenters holding batches of Pinot Noir grapes. These fermenters are much smaller than those of other wineries I've visited, especially ones specializing in Cabernet. As Tony explains, it's difficult to make high-quality Pinot Noir on a large scale. The grape needs a lot of attention. They're clearly getting it here.

"Pinot Noir is a weird bird compared to Cabernet," he says, showing me around the bubbling fermenting tanks in the converted barn. "It's missing a pigment and has seeds that are highly tannic. If you throw a Cabernet grape against the wall, it will bounce off. If you throw a Pinot grape against the wall, it will splat."

A delicate grape, Pinot Noir requires more care than Cabernet, which is easy to extract. You stem and crush the grapes and pump them into big vats for fermentation. That's the wrong approach with Pinot.

"If you want to make a Pinot that has good color but is not coarse with tannin, you have to first cold soak whole clusters," he says. "Ferment Pinot with whole berries very gently, trying not to crush the grapes and extract more seed tannin."

These fermenters hold two tons of Pinot Noir. Cabernet fermenters can hold twenty tons or more. The smaller fermenters proved a big improvement in making Oregon Pinot.

"The small vats have a higher surface to volume ratio," he says, showing off the fermenter. "Small vats allow color extraction."

It took him years of experimentation to work this out. After graduating from college with a degree in philosophy, Tony moved to Napa in 1975 to make wine. He landed at Stag's Leap Wine Cellars, scrubbing tanks for Warren Winiarski. In 1982 he became winemaker at Spottswoode. He went on to become a consultant, working with such prestigious houses as Dalla Valle, Viader, Niebaum-Coppola, Araujo, and Shafer. While mastering the intricacies of Cabernet, he fell in love with the finicky, challenging, expressive Pinot Noir grape. In 1982 he launched Etude, his label specializing in California Pinot Noir.

"My first brand adventure was Etude," he says. "I figured it would take ten vintages to better understand how to make Pinot. There was a lot going on in the field. I was trying to get more concentration in the wine."

By thinning the crop, Tony reduced the yield from four to five tons per acre to two tons. The concentration and quality increased. Tony began to pay growers by the acre, not by the pound, for grapes, allowing them to make a decent profit with the lower yields.

By the mid-1990s, he decided to move to Oregon with his new wife, Michelle, also an Oregon native. They purchased Mineral Spring Ranch in 2001 and never looked back. Michelle died of cancer in 2019, but Tony continues to pursue their vision of a biodynamic farm and winery. Their children, Olivia and Anton, love the property but are not currently planning to take it over. At seventy, Tony remains at the helm, walking the vineyards and making wine for as long as he can. He shows few signs of slowing down.

After touring the winery, we hop in his gray Audi station wagon to check on the grapes on Ribbon Ridge. It's farming that really makes the difference for him. He says that every vineyard has a voice; you just have to listen.

"The real effort is in farming," he says. "We're looking for flavor concentration, maturity, and quality of fruit."

As we walk through the Ribbon Ridge vineyard, Tony stops to taste the grapes, making sure that they're ready to pick.

"These are Pommard clones," he says, pointing to the grapes. "They have little dimples in the skin when they're ready. These are twenty-four Brix and will produce a wine of 13 percent alcohol."

He uses the Brix measure as a rough gauge of the harvest, with the art and judgment of the winemaker in the field determining the final decision.

That's what he's doing today. The grapes are black in color, shiny, and sweet to the taste and seem ready to pick.

"This is a UC Davis selection," he says, tasting it and explaining the characteristics of the various clones in the vineyard. He speaks clearly and succinctly, as if giving a graduate-level seminar in viticulture, but there's nothing abstract about it. It's very practical and down to earth.

"David Lett, one of the earliest establishers of Pinot vines here, chose them," Tony says. "It turns out they're really good clones. We didn't always appreciate how to farm them. But if you can keep the crop down, the wines can be terrific."

He tastes again and steps out of the way as a picker runs by with five-gallon plastic buckets. "In the old days, if the crop was left to its own devices and we didn't thin, the wines would be lacking in concentration and compromised by the late-season weather half the time."

He gestures with his hands to emphasize the point. "Why are the odds so inconsistent in Oregon?" he says of the old days. "Why do you get only three great years out of ten, and two years you have to write off as mediocre and the others are in between?"

This observation corresponds with my sense of Oregon Pinot over the years. While I've enjoyed the wines, I sometimes found them lacking in concentration.

"Oregon has a limited climate," he says. "The wine grapes get only so many days of sun, which limits the yield. We're looking for two tons of grapes per acre rather than five. The breakthrough came with our Chardonnay."

It's been a process to figure out the best way to grow grapes in the Willamette Valley. Limiting yields to two to two and a half tons per acre proved key. "Pinot Noir is sensitive to crop levels," he says. "The quality diminishes with larger yields. You never get really interesting wines with larger yields."

With the decrease in yields and increase in quality, Oregon Pinot prices have risen steeply; it's difficult to get a good Oregon Pinot Noir for less than forty dollars.

"It's tough to make a ten-dollar Pinot," Tony says. "The prices of Pinot have come up enough to make the economics work. We're in the fine wine business."

We keep walking. "Today, we're making really consistent wines regardless of the vintage," he says, tasting another grape. "You could say part of that is climate, but it's also the root stock, which helps control vigor and enhances maturity, and the discipline of crop thinning to get quality."

There's a lot to ponder. I don't have the same level of control over my grapes, but these ideas are worth considering. I taste along with him, trying to get a sense of what he's talking about. The fruit seems quite ripe. Fat bunches of black-purple grapes hang from the vines, ripe, shiny, and healthy.

Tony keeps walking, tasting along the way. He picks a grape and holds it up.

"As far as when to pick, the mature grape comes off the stem very easily," he says. "It leaves a little flesh behind. It feels like velvet in your hand. It has a little tenderness to it. If they were picked a week ago, they would be firm and bouncy."

I pluck a grape and test its soft surface tension. It's very velvety. I pop it in my mouth and taste it—delicious.

"You're looking for black velvet quality of skins," he says. "The grapes come off the vines ready to give themselves up."

He points to another cluster. "They still accumulate color in the last week of harvest. Waiting those extra days, if the weather allows, gives you an extra measure of concentration. The more tender they are the easier they'll give up this color. And you don't have to work very hard to do it. If you work hard, you end up squeezing out the seeds."

He squeezes a grape to demonstrate. "The seeds in any grape are bitter, and in Pinot Noir they're much more tannic than the seeds of other red grapes. That's a rock and a hard place for a winemaker who's trying to get all this color and doesn't want that tannin. You get a thin, sometimes bitter, astringent Pinot Noir because it's been handled so roughly."

I've had more than one of such wines. In fact, it's not a bad description of the Pinot I made early on in my wine-making career.

A tractor passes by. "Watch out," he warns.

I duck out of the way.

The roar of the tractor drowns out our conversation. We keep going. The vineyards slope downward, radiant yellow in the fall, with clear skies and the tang of smoke in the air. He calls out to the vineyard manager who supervises the picking. The guy gives him a thumbs up. It's going well.

"There's a unique sense of place here," Tony says. "The wines taste different, which makes it enjoyable."

He's always trying to learn more about the fruit, trying to strike a balance. "If you go after color, you might make the wine more tannic," he says. "It's a balance of trade-offs."

After visiting Ribbon Ridge, we walk back to his car and drive back to the 240-acre Mineral Spring Ranch, where he raises chickens, pigs, cattle, and vegetable gardens. Its biodynamic vineyard is certified organic.

"It's marine sediment and sandstone," he says. "It's dry-farmed in most years, which gives it a sense of place. The goal is supple, sexy Pinot with structure to age."

He considers the property part of an ecosystem, with vines surrounded by grains and other plants. "In college I studied philosophy, so I have some skepticism about biodynamics, but it's prescient in its characterization of what's going on. It is an intuitive understanding of an infinitely complex system. I think our understanding of 'organic' and 'biodynamic' will keep evolving as we get a better understanding of best practices, trying to optimize nutrition and [the] flavor of plants."

He approaches biodynamics as a whole-farm philosophy. Everything on the property has to be compliant: vines, grasslands, trees, orchards, and animals. This sometimes involves finding new solutions to old problems. When birds hit the vineyard like something out of the Alfred Hitchcock movie *The Birds*, gobbling up precious grapes, Tony came up with a unique solution: He hired a falconer to discourage other birds from visiting the vineyards.

"We're looking to nature for means of interaction," he says. "It's bird control employing nature's hierarchy. It works fabulously well. We're tuning into nature's mechanisms."

He emphasizes that organic and biodynamic practices can yield wines of the highest quality. "You get depth of character with these practices," he says. "Like old vines, they add another dimension. We also use spontaneous fermentation with naturally occurring yeasts, not cultured yeast."

All these practices and evolution have led to big leaps in quality, with his latest vintage as an exemplar.

"We've had some of our better vintages in the last two decades," he says reverently. "I can't emphasize the gratitude I have for making standout wines."

Tony walks me over to the tasting room, an elegant, modern barnlike building with polished stone floors, high ceilings, wood-paneled walls, and territorial views over the Willamette Valley below. We shake hands, and he takes his leave. He still has to check on other fruit. A winegrower's work is never done, especially at the harvest.

Hallie Whyte, the managing director, greets me and brings out some wines to taste. A lively, dark-haired woman, she grew up in the Willamette Valley and worked in local restaurants, which led her to join the growing wine industry.

"It's been a roller coaster of a season," she says, setting out a sampling of Soter wines. "Everyone's anxious until the fruit comes off the vines."

I'm a fan of Soter sparkling wines but have yet to taste their estate wines. She first pours the 2019 Soter Vineyards Estates Chardonnay, a delightful blend of several vineyards with a bright apple aroma and a snappy, flinty finish. When I mention its resemblance to a white Burgundy, Hallie nods but emphasizes how Oregon has come into its own as a region. Once upon a time, Oregon depended on the Burgundy comparison. Now, not so much.

"We've stopped comparing ourselves to Burgundy," she says. "Wine is complicated."

The next wines are also complicated in a layered, balanced way. We taste the 2019 Estate Pinot Noir, a rich, round blend. Then we move to the 2020 Mineral Springs Ranch, a single-vineyard wine that trills an exquisite melody in my mouth. I look back toward the vineyard that produced it and over the Willamette Valley in the distance. Right here, right now, this wine is a perfect expression of this place.

14

Ken Wright Visit

Wine as Fundamentally Spiritual

In the darkness before dawn, fog engulfs downtown Carlton, a small town in the Willamette Valley. The fog swirls around the sprawling complex of Ken Wright Cellars, infiltrating the crushing facility and obscuring the tasting room and business offices.

Out of the fog comes a forklift. Ken Wright, owner of Ken Wright Cellars, speeds the forklift toward the parking lot where I'm standing. He brakes, jumps out, shakes my hand, and welcomes me to the winery.

A former high school wrestling champion, Ken is fit and squarely built with an easy grin and a magnificent mane of gray hair. He wears a brown plaid shirt, dungarees, boots, and a baseball cap. It's early, but he's buzzing with energy as he works to bring in the grapes and close out the harvest.

This year has been extraordinarily challenging. An April freeze damaged many of the vines' buds, but careful management allowed him to salvage the crop. The flowering of the vines occurred in July, rather than June, with the warmer, drier weather yielding more fertility than in his previous years in Oregon. A long dry summer allowed the grapes to ripen fully. Now storms have lined up off the Pacific Coast, threatening to inundate the fruit. Ken wants all of it picked pronto.

"There's no recipe," he says, shaking his head. "Every harvest is different."

He should know. Moving to Oregon in 1986, he made his first Pinot Noir for Panther Creek Cellars. Since then he has become one of the masters of Oregon Pinot, innovating by using sorting lines to cull grapes, dry ice to cool grapes before fermentation, and paying growers by the acre instead of the ton to enhance fruit quality. But perhaps his most lasting legacy remains the creation of six sub-appellations in the northern Willamette Valley: Yamhill-

Carlton, Chehalem Mountains, Ribbon Ridge, Dundee Hills, McMinnville and Eola-Amity Hills.

Ken discovered the nuances of the regions while making wines and tasting those of others. He suspected the subregions defined the wines' character. Building off this, he pioneered single-vineyard Pinot Noir and campaigned to get his fellow vintners on board. Not all were enthusiastic, but Ken's energy and persuasiveness won out in the end. In 2005 the subregions were created, allowing vintners to explore all the nuances of the territory and wine lovers to enjoy the stunning diversity of wines from this place.

He's invited me to witness these innovations in person, arranging for his assistant, J'Aime du Mauriée, to serve as my guide. He's promised to answer my questions, but getting Ken to sit down, especially during the harvest, is nearly impossible.

He jumps back on the forklift while J'Aime, a slim, dark-haired woman with an encyclopedic knowledge of the business, steers me toward the company van. We drive south to Carter Vineyards, one of the sites where Ken first got an inkling of the diversity of Willamette Valley geology and convinced him of the need for subregions.

On the drive down, J'Aime fills me in on the Carter vineyard, which is located in the Eola-Amity Hills AVA and sits on volcanic soils that dry quickly, encouraging early ripening. The wines exude aromas of black cherry, plum, and cassis. With acidity levels higher than those of other growing areas, the wines possess structure and ageability.

We pass groves of hazelnut trees as the road winds through the countryside. The Willamette Valley is rich agricultural land with wheat, oats, hay, vegetables, and other fruits cultivated on the valley floor while vineyards line the hillsides. We turn onto a gravel road, passing farms and rolling hills. J'Aime navigates around potholes, crosses a field, and parks at the base of the Carter Vineyard. While the surrounding farms are quiet and peaceful, the vineyard is bustling with activity. Pickers jog by with buckets of fruit. A tractor loads bins of grapes onto a semitruck for delivery to Ken Wright Cellars.

Carter and nearby Canary Hill Vineyard were both planted in 1983, but their fruit is remarkably different. These sites got Ken thinking about the diversity of geology in the valley. A west-facing site, Carter gets the late sun. The vines adapt by thickening the grape skins, yielding more tannin and more structure.

"There's a huge difference between Canary and Carter," Ken says of the vineyards. "They are both volcanic so they are more fruit driven, but Canary is more feminine and more pretty, with a lot of soprano in that wine. Carter is dark, deep, dense, more masculine, much of that due to being west facing. Carter has the ability to age gracefully. This is a wine to lay down for a special occasion."

I wonder how the vineyard will reveal this. J'Aime leads the way into the vines. There's dew on the grass, a chill in the air, and mist circulating through the vines. Seeing my breath, I zip up my coat and enter the rows. Fat black clusters hang from the vines, the berries dark purple, glistening with dew, and ready for picking. I pop one into my mouth and savor the rich black cherry flavors.

The bright orange ball of the sun appears, piercing the fog. I walk back to where Seth Miller, the tall, dark-haired, animated vineyard manager, oversees the pickers who move quickly through the rows, laughing, joking, clipping clusters, and dropping them into five-gallon white plastic buckets. They empty the buckets into large plastic totes. Seth removes the occasional leaf from the grapes, but the fruit is very clean. A tractor roars by, dragging bins to a semitruck, which will take the fruit back to Ken Wright.

It's all very efficient and purposeful, with everyone smiling and taking pleasure in the harvest, relieved the fruit is so fantastic after a very challenging year. Surrounded by all this activity, it's hard not to get caught up in the excitement of the harvest. It reminds me of why I started making wine.

After visiting Carter, J'Aime runs me back to the winery to help sort the Carter fruit. I've done this at other wineries, but it's always instructive to see each house's approach.

Back at the winery, Ken whizzes past in the forklift, bringing in bins of Carter fruit. "Look before you walk," he says, warning me out of the way. "Anything can happen."

I nod and step aside as he ferries the bin to the hopper, lifts it up, tips it on its side, and dumps its contents into the stemmer/crusher.

"I have a question . . . ," I say, but he's off, speeding outside to pick up another load of grapes.

I find a box of surgical gloves, put them on, and find my way to the sorting line. At harvesttime, it's all hands on deck. Volunteers and winery employees line either side of the sorting table, which shakes and rattles as the fruit comes down. We remove leaves and underripe berries as clusters move toward the stemmer/crusher.

The fruit is cold from the morning pick. I work my way through the clusters, discarding leaves, stems, and bugs. The fruit's pretty clean, but I want it perfect.

When things slow down, I taste the grapes. Their sweet, rich, dark fruit flavors are like those I sampled back at the vineyard, with a balanced acidity that will make for a wine that ages well.

The next container needs more culling. Some clusters show botrytis, a kind of gray mold that often occurs near the harvest, especially in dry years such as this one. Botrytis is essential to producing sweet wines such as Château d'Yquem of Sauternes, France, but not essential for Pinot Noir.

"Is this cluster okay?" I ask Ken's son, Carson Wright, who is the general sales manager. I hold up a cluster showing botrytis.

"You can toss it," he says.

I drop it into the discard bin. I get to assume some responsibility here. The operation is congenial with clear directions and lots of opportunities to pitch in. When juice accumulates in a bucket below the sorting line, a volunteer shows me how to empty it into the fermenter. That becomes my job. Now I have two jobs!

Taking a break from the sorting line, I move to the fermenting room. It's wide and spacious with high ceilings and rows of large plastic fermenting tanks. It reeks of fermenting yeast and fruit, with large fans blowing off the CO_2 produced by the fermentation. My task is to measure the temperature of the ferment and the residual sugar, and smell the must. If one of the tanks ferments too quickly or too slowly or smells off, the team can quickly correct it.

Winery ambassador Ivory McLaughlin, a young blond woman wearing a pile vest and checkered shirt, shows me how to measure for sugar and temperature. I need a few tries to get it right; then I work my way through the bins.

I report the numbers to Joylin Kent, the lively, dark-haired wine club manager, who records the numbers on a clipboard. "Point eight, ninety degrees," I call out. "Point five, eighty-eight degrees. One point two, eighty-nine degrees."

We move methodically down the fermenting tanks. It's not hard work, but it's repetitive and requires focus.

After checking the sugar and temperature readings, we measure out Superfood, which contains DAP (diammonium phosphate); the yeast nutrient helps with the fermentation. We also use this approach at Les Copains. It's reassuring to note that our protocol follows that of a renowned house like Ken Wright Cellars.

By the end of the shift, I'm drenched with sticky grape juice. Grape seeds cling to my pants. An earwig is crawling up the back of my shirt, but I don't care. I've helped out during the biggest day of the year. When I wash my hands and get ready to leave, Ken is still driving the forklift, bringing in the fruit.

Later that evening, Ken invites me to dinner at Earth & Sea, an outstanding restaurant a block from the winery. I arrange to sit next to him to ask my questions. He's changed from his harvest clothes into a light dress shirt, casual pants, and jacket. I also get to meet his wife, Karen, a calm, dark-haired woman who designed much of the winery.

Ken orders oysters, fish, steak, and other dishes that we share family style. He chooses wine from local producers he admires such as Mike Etzel and Tony Soter. It's a revelation to taste these wines with the outstanding cuisine, always my favorite it way to enjoy them, and brings my trip full circle.

"How did the harvest go?" I ask.

"It went from disastrous to miraculous," he says excitedly. "It was like, 'Buckle your seatbelts. Stow your tray tables and carry-ons.' In April we had a cold front, not just a frost. Four days in April fell below thirty-two degrees. It was my forty-fifth harvest, and I'd never seen such cold damage. The literature suggested that recovery was impossible. It looked like we might get twenty-five percent of the crop."

"How did you turn it around?" I ask.

"We used the volunteer shoots of the vine that normally would be rubbed off, hoping they would produce enough fruit to offset the initial cold damage," he says. "The 90 to 95 percent fertility of the flowering was the first miracle. It was helped by the period of flowering happening in July with warm temperatures and clear skies rather than in June, which is normally cool and rainy in Oregon. But the development was late, so we needed a longer season to mature the fruit. Guess what didn't happen? It didn't start raining as it usually does in the fall. It was a miracle year in Oregon."

"The fruit looks fantastic," I said. "It looks fully ripened."

He smiles as if in relief.

"And I noticed how carefully you treated the fruit at the winery," I added.

He nods enthusiastically. "The key thing with Pinot is you need small lots," he says. "Cold soaking of those small lots is the best approach. Then you get color and flavor without the bitter and astringent tannins, which are in the seeds."

This approach seems essential to getting the character of vineyards such as Carter's into the wine. It is very similar to the approach of Tony Soter and Mike Etzel. Many now agree on how to make the best Pinot Noir in Oregon, with vintners like Ken, Tony, and Mike leading the charge. They've cracked the code.

"We were the first to focus on single vineyards," he says. "The beauty of Pinot Noir is that it connects you to place. It's amazing."

Ken's mantra from early on has been "Listen to the vineyards. Be grateful for what they are giving you."

"It's deeper than grand cru," he says, bringing his hands together. "I was at Domaine de la Romanée-Conti and met director Aubert de Villaine. He's one of the humblest people ever. He's an amazing person and amazing farmer."

Ken learned from de Villaine to listen to the vineyard. Don't impose your ideas, theories, and practices on a place. Learn how to express it in your wines.

"How do I become invisible and transparent?" he asks. "It takes understanding that greatness happens when the plant and place are perfectly matched. Our fruit is world class. We have the goods. We have the gift, but that comes with the burden of protecting those qualities that come from the field."

This emphasis on expressing *place* explains his push for sub-appellations within the Willamette Valley.

"We needed to identify world-class plots," he says, leaning forward in his chair. "We needed to create a road map. I was trying fruit from all over the valley. I realized I wasn't aware of all the differences. In 2000 we formed six subregions based on geology. We met and worked out most of the issues."

In 2005 the AVAs became official. Since then, Ken has doubled down on his approach. He makes single-vineyard Pinot Noir from many sites in the valley, including Carter and other vineyards.

"It's about honoring great places," he says, bringing his hands together as if in supplication. "It's a next-level thing. You feel that connection."

I certainly feel that connection at Ken Wright, Beaux Freres, and Soter. Oregon Pinot seems to have taken a quantum leap forward in the last twenty years, following the lead of these and other Willamette winemakers.

"Pinot is a blank canvas," Ken says, raising his hands. "It reveals place. It's amazing to have a wine that takes you there. Pinot Noir is unequaled in its ability to take you there. It makes wine fundamentally spiritual. It's not about me. World class is about plant and place—not people. If you're blessed to work with world-class fruit, it's an amazing gift. This is what the place gives."

The next morning, I rise before dawn, ready to head back to Seattle. The fog has crept in again, but the weather is holding—at least for today. Ken is already up, speeding the forklift around the winery and working hard to honor the gift of the fruit and the place.

Fig. 1. Les Copains founders Tom Remmers (*left*) and Nick O'Connell pick up grapes at Ciel du Cheval Vineyard at the beginning of the harvest. Buying grapes from this storied vineyard makes all the difference in the quality of the finished wine. Photo by Kade Casciato.

Fig. 2. Jim Holmes at Ciel du Cheval Vineyard. "You can't get any closer to terroir than this, studying the soil and stones that make up the vineyard—all key to understanding how to make great wine." Photo by Nicholas O'Connell.

Fig. 3. Bob Betz sorting grapes at Betz Family Winery. "Wine making is about pleasure," he says. "That's my philosophy." Photo by Nicholas O'Connell.

Fig. 4. Sorting grapes at Betz Family Winery. I shove my hands into the grapes, which are cool to the touch, squishy, sticky with sugar. Working on the sorting line you really get to feel the texture of the fruit. Photo by Nicholas O'Connell.

Fig. 5. Chris Upchurch of DeLille Cellars and Upchurch Vineyard. "I think science is overrated in wine making . . . ," he says. "Wine making is an art and craft." Photo by Kristina Muller-Eberhard.

Fig. 6. Brian Mackey of Chateau Ste. Michelle. "Wine makers love the harvest," he says. "Christmas has nothing on the harvest. . . . Harvest is the key to making great wine." Photo by Nicholas O'Connell.

Fig. 7. Carter Vineyard at dawn. The vineyard is bustling with activity. Pickers jog by with buckets of fruit. A tractor loads bins of grapes onto a semitruck for delivery to Ken Wright Cellars. Photo by Nicholas O'Connell.

Fig. 8. Carter Vineyard grapes. Fat black clusters hang from the vines, the berries dark purple, glistening with dew and ready for picking. I pop one into my mouth and savor the rich black cherry flavors. Photo by Nicholas O'Connell.

Fig. 9. Mikey Etzel with fermenting Pinot Noir at Beaux Frères Winery in Oregon. "There's no recipe. No secret. We don't compromise on what we feel is right. This is our life. We eat, breath, shit, and bleed this place." Photo by Nicholas O'Connell.

Fig. 10. Tools of the trade. Pruning shears at Beaux Frères Winery. Photo by Nicholas O'Connell.

Fig. 11. Tony Soter overseeing the harvest at Soter. "There's a unique sense of place here," he says. "The wines taste different, which makes it enjoyable." Photo by Nicholas O'Connell.

Fig. 12. Nicholas O'Connell (*left*) and Ken Wright at Ken Wright Cellars at harvest. "It's about honoring great places . . . ," he says. "You feel that connection." Photo by Nicholas O'Connell.

Fig. 13. Ivory McLaughlin (*left*) and Joylin Kent checking fermentation tanks at Ken Wright Cellars. I'm drenched with sticky grape juice. Grape seeds cling to my pants. An earwig is crawling up the back of my shirt, but I don't care. I've helped out during the biggest day of the year. Photo by Nicholas O'Connell.

Fig. 14. Warren Winiarski inspecting the fruit at Arcadia Vineyard. "You have to listen to the grapes," he says. "You have to go back to the vineyard." Photo by Nicholas O'Connell.

Fig. 15. Michael Silacci pours the magical 2010 sampling at Opus One Winery. "Taste is smell," he says, taking a sip. "I'm getting dark bitter chocolate with a long-lasting finish like church bells at the Vatican." Photo by Nicholas O'Connell.

Fig. 16. Chris Howell paying homage to the harvest at Cain Vineyard and Winery. "Each wine has to have its own identity, derived primarily from the vineyard. Our job is to get to know the fruit and allow that expression to come through." Photo by Nicholas O'Connell.

Fig. 17. The connoisseur: Dan McCarthy of McCarthy & Schiering Wine Merchants. It's a clean, well-lighted place with a tang of yeast, a hint of wood, and a long, lingering effervescent finish. Photo by Janet Hunter.

Fig. 18. Bottling day at Les Copains Winery. Once volunteers get the hang of it, the bottling line starts cranking. Everyone joins in to make it happen. Photo by Nicholas O'Connell.

PART 4

My Wine Journey

15

Falling in Love with Wine

I fell in love with wine in Nantes, France, during a junior-year abroad program. While on a field trip to see the region's architecture, we stopped at a local winery. The winery building was constructed of cinderblock, with a dimly lit tasting room that served the local Muscadet, a wine I'd never heard of.

I approached the bar for a taste of wine. One of the *ouvriers* (workers) handed me a glass and poured me a splash. I was not yet twenty-one, so it was a thrill to be served alcohol, even though I'd been drinking wine for the three months I'd traveled in Europe prior to the class. The ouvrier demonstrated how to swirl the wine around in the glass to increase the aroma.

I did this and took a sip. The wine puckered my mouth; it was crisp, dry, and slightly acidic. I then tasted it with slices of bread and goat cheese.

The other students joined in, eager to taste but unsure of the rules. The workers gave everyone a taste and talked about what they were drinking. It was not like a fraternity party where everyone tried to get hammered and blow beer out their noses. It was lively and convivial but civilized in a non-snooty and appealing way. The wine we tasted, Muscadet, a white wine from the region, would never sell for hundreds of dollars, but that didn't matter. It was their local wine, and the ouvriers were proud of it.

The workers had red cheeks and blue overalls, and joked around with each other. It was the end of their day caring for the barrels of wine stacked in the cellar, and they all shared a glass too. It seemed a cherished part of their life, and they were letting me in on it. They told me about the wine and the vineyard it came from. Despite my limited French, I began grappling with terms such as "terroir" and "fermentation."

A warm glow settled over me. During the trip I'd been homesick for family and friends in Seattle and wanted to break out of the circle of the program's fifty American students, but I didn't speak French well. Restaurants and bars closed early in Nantes. Making friends outside of our group of American college students had been difficult.

I sipped the slightly fizzy new wine, chatted with the workers, and got pleasantly buzzed. While the other students scrunched their faces at the wine's tartness, I talked with the workers, enjoying being invited into their circle. I want to learn more about it. I liked its tartness and subtle melony flavor. They described all this with gestures and complicated explanations, most of which went right over my head. I nodded a lot and got offered a refill.

By the time we got back on the bus, I was slightly tipsy, not so much on the alcohol as on the whole experience. The French are often regarded as a formal, sometimes rude, people. But the workers were warm, funny, and approachable, generously introducing me to one of their favorite rituals.

After that tasting, I was hooked. I had little money but scraped together enough for a bottle of wine once a week. After Sunday Mass inside the Quonset hut of a nearby Catholic church, I bought a dozen local oysters that were sold in plastic tubs, two for a franc. Coming from Seattle, I had a taste for oysters, but these came from the chilly waters of Brittany. They were a revelation: cold, briny, with a satisfying crunch. I'd also buy a loaf of bread and a bottle of Gros Plant, a cheap local wine. This became my Sunday lunch.

The wine was not refined, but its flavor began to change my palate, just as living in France changed my larger perspective. I no longer wanted something sweet. The Gros Plant and the occasional bottle of Muscadet provided a stepping stone into the magical world of wine.

On weekends, I'd use my Eurail pass to visit Germany, Italy, Spain, and other parts of Europe, often sleeping on the train or in the train station to save money. I'd always buy a bottle of wine for the journey. It was cheaper than bottled water and provided a window into the culture I was experiencing: the pine tar taste of Retsina in Greece, the grapefruit tang of a Riesling in Germany, the earthy, tobacco taste of a Bordeaux. A glass or two would help me sleep, even under the seats of the train.

One trip, my friend Kurt Eisen and I arrived late to a youth hostel in Rome. We were famished and searched for a grocery store or cheap restaurant to get something to eat. All were closed. We wandered down the street, looking for a place that was open. We spotted a well-lit restaurant a block away with white tablecloths, fine china, and waiters wearing black dinner jackets. We were so hungry we didn't bother to look at the prices.

The waiter found a table for us and brought us each a large bottle of beer and followed with a bottle of Chianti, red peppers stuffed with rice and beef, pasta with fungi, and a *bistecca alla fiorentina*, which was one of the largest steaks I'd ever seen. It was a whirlwind of new tastes, textures, and pleasures, a visit to a world almost unimaginable for someone from Seattle of that era. The bill was shocking, many times the amount we'd allocated for meals, but we managed to pay it and stumble back to the youth hostel.

That memory stuck with me. I'd walk by nice restaurants in Paris and look in the window, mesmerized by the smells wafting out, hearing the pop of a wine cork, seeing the waiters bustling back and forth from the kitchen with gleaming plates of duck confit for the well-healed patrons. One look at the menu prices sent me packing.

Instead, I'd head for the neighborhood *boucherie* (butchershop) or *épicerie* (grocery) to buy some pâté, local cheese, a baguette, and a bottle of the local wine. Then I'd take my *picnique* to the local park, where I'd sit on the bench, watch the sun set, and enjoy my cheap but delicious dinner.

When I wasn't traveling, I'd dine with my hosts, the Cochois family. Madame was an excellent cook, providing us with breakfasts, dinner twice a week, and a four-hour-long Sunday meal with the extended family. The Cochoises' grown children and their husbands and would join us, giving me a great time to practice my French and to enjoy such delicacies as sausage, rabbit, coq au vin, and sweetbreads in cream sauce en croûte.

Madame once served a dish that looked like breaded veal but turned out to be breaded cow brains. I took one bite and gagged on the tracery of tiny blood vessels in the gelatinous dish. I ate green beans and bread instead. But I usually loved her food so much that I earned the nickname *la poubelle* (garbage can), because I was always eager to finish up the leftovers.

To complement the meal, they served a local wine such as Muscadet or occasionally a red Burgundy or Bordeaux. Afterward, Monsieur would take me aside for a tasting of digestifs and liqueurs.

One time he held up a glass of Cointreau. "Pour les femmes," he said, taking a taste before pouring me a splash.

I sniffed it, enjoying the smell of oranges, and tasted the deep, sweet orange flavor. "Très bien," I said, setting down the glass.

"Comme ci comme ça," he said, wagging his shoulders to indicate he didn't like the sweet Cointreau.

Then he brought out a bottle of Calvados. "Pour les hommes," he said, nodding knowingly and poured me a splash.

I inhaled deeply, nearly gagged, and then choked down the liquid fire, which burned my throat as if I'd swallowed a white-hot apple. "Formidable!" I managed to say.

"Bien sûr," he said, clapping me on the back. He had four daughters and no sons, and seemed to enjoy having me around as an occasional drinking companion.

My father enjoyed gin and vodka, but had never shared them with me. That was probably because I was not yet twenty-one and because of the awkwardness of the situation in the United States, where many considered alcoholic drinks as more of a drug than an integral part of a meal.

Though I went to France to study language, history, and culture, I learned more from my daily interactions with Monsieur and the rest of the Cochois family. I returned to the States eager to incorporate what I'd learned into my life back home.

After graduating from college, I worked as a journalist for newspapers in Washington State. The state's wine industry was just taking off, with Ste. Michelle Johannisberg Riesling showing up on the shelves. I sprang for a bottle, intrigued about northwest wine. It was a sweet, appealing wine with a measure of tartness to keep it in balance.

The newspaper jobs didn't pay much, but I lived cheaply and managed to save my pennies for regular trips to Europe to indulge my growing wine habit. I took the train from Paris to Beaune, a walled town and the capital of the Burgundy region. As the train approached the town, I took in the spires of Gothic cathedrals; the colorful, geometric-patterned roof of the fifteenth-century Hôtel-Dieu; and the green undulating vineyards flowing down the nearby hills.

After finding a youth hostel, I headed for the Marché aux Vins, an underground wine cellar serving local wines. I hoped it would be a cheap and enjoyable introduction to Burgundian wines. For ten euros, I could taste dozens of local wines. I paid the fee, received my tastevin (metal cup), and entered the cellar. My God, it was almost too good to be true—a self-guided introduction to the wines of Burgundy! I quickly helped myself to a white wine made from Aligoté grapes. It was tart, lemony, and quite acidic, not to my taste. I spat it into the *crachoir* (spit bucket).

As I moved farther into the cave, the world of Burgundian wine opened up. I recognized the names of the famous villages where the wines came from: Côte de Beaune, Côte de Nuits, Santenay, Chablis, Pommard, and Meursault. I had wanted to taste these wines before but had shrunk from the cost. Now I could sip them at my leisure.

The whites were made mostly from Chardonnay grapes, but the characteristics were very different, providing an early education in the role of terroir. I loved the crisp minerality and oyster shell taste of Chablis. I admired the green-gold color of a Meursault and its whiff of hazelnuts. I forced myself to taste, spit out the excess, and move to the next one. I had a lot of wine to sample and didn't want to dull my senses.

The reds were remarkable, tasting of cherries and smelling of earth. I sipped a lively Côte de Beaune; a rich, earthy Côte de Nuits; and a perfumed Volnay. I was walking on air. I longed to taste the premier cru wines, such as Puligny-Montrachet, Gevrey-Chambertin, Chambolle-Musigny, and Domaine de la Romanée-Conti—one of the most expensive wines in the world—but such wines were not available at ten euros.

Near the end of the tour, I struck up a conversation with an American couple. They were from the Midwest, new to wine, and eager to try more expensive bottles. The husband sported a natty polo shirt, white shorts, and a Rolex watch. The wife was blonde and vivacious, and wore a colorful print dress and a large diamond ring on her finger.

"We want to taste the good stuff," the husband said.

"Me, too!" I said.

The husband introduced himself to a middle-aged salesman. The salesman shook hands with him but had trouble speaking English. I stepped in to translate.

"Nous voudrions goûter les meilleures vins de Bourgogne," I said, explaining we'd like to taste some of the best Burgundian wine.

The Marché aux Vins shrewdly put the best wines at the end of the tour, figuring that if people drank along the way they'd be buzzed and more likely to whip out their credit cards for a big splurge at the finale.

The salesman looked us over. The couple from the Midwest clearly had money to burn and seemed worth the cost of opening some of the most expensive bottles, but my hiking shorts, sandals, and T-shirt didn't suggest the kind of person who would purchase multiple cases of Domaine de la Romanée-Conti. He hesitated but decided we were a package deal. He took a key from his pocket and opened a dusty old cabinet.

He took out three bottles and set them atop a wine barrel. Pulling out a corkscrew, he opened a bottle of Puligny-Montrachet, Chambolle-Musigny, and Gevrey-Chambertin. I couldn't believe my luck.

"Formidable," I said.

He smiled politely. Then he poured a splash of the Puligny into the wife's glass.

"It's good," she said.

He poured a splash for the husband.

"Not bad," said the husband when he tasted.

He poured me a taste. I swirled it around in the glass, smelled the intense citrus, and sipped the luscious nectar. I'd never tasted a wine so rich and concentrated. It made other white wines seem like soda pop.

"Fantastique!" I said.

He smiled enthusiastically and moved on to the reds. It went pretty much the same way. The husband and wife obviously liked the wine, but I *loved* it. I seemed to have a sense they lacked. The Gevrey-Chambertin gushed with an opulent crushed raspberry taste mixed with rose petals and minerals, while the Chambolle-Musigny played an exquisite melody in my mouth. I tasted and retasted it, wanting the music to last.

The husband whipped out his credit card. The salesman started filling cases. I headed out of the cellar, thrilled to have tasted so many outstanding wines, and having discovered that even if I couldn't afford such wines, I could appreciate them. I was a taster.

Several years later, I visited Europe with my girlfriend Clara Menhuin-Hauser. She was charming and witty, and loved the finer things in life, including

Champagne. After visiting her family in London, we traveled to France to tour the Epernay region. We spent a day visiting the various Champagne houses, tasting and spitting. At lunch, we had an impromptu picnic in the park, a higher-end version of the ones I'd enjoyed as a student. We dug into a whole roast chicken and shared a bottle of bubbly. I got caught up in the moment and drank too much.

Our last tour took place at a smaller Champagne house. The woman leading the tour was short and blonde, with a flirtatious smile and curvaceous figure. After the tour, she poured us samples of their wine. Each was an education in its own right: a tart, earthy bottle made from mainly Pinot Noir grapes and one from Chardonnay grapes that smelled like bread dough. For the finale, she poured a vintage pink Champagne, tart, fruity, and effervescent. It was fantastic, light years beyond my price range, and a hint of what a great wine could be. I tasted and retasted it, enjoying it so much I didn't realize the effect it was having on me.

"What do you think?" the tour guide said, holding up the bottle.

"Well," I said, trying not to stare at her chest, "it's a really full-breasted wine."

The tour guide blushed.

Clara laughed and elbowed me in the ribs. It was an apt description of the wine: full, ripe, and sensual, just like the woman pouring it. That marked the beginning of my career as a wine writer.

While pursuing a graduate degree in English, I had little money to indulge my wine habit, so I attended lectures—not for the talk but for the wine and cheese afterward. The University of Washington's English Department served wine and cheese at these readings, usually a Mondavi Cabernet and Kendall-Jackson Chardonnay with chalky cheddar and bland Monterey Jack cheese on toothpicks. It wasn't great, but I didn't have to pay for it.

When I wanted something better, I'd attend talks by speakers from the business school. Before the speaker finished, I'd head to the lobby to gorge on baked Brie and sliced sausages washed down with Chateau Ste. Michelle Cabernet. The higher the speaker's profile the better the food and drink.

In the fall, Clara and I took an RV trip to California. Trying to park the RV in San Francisco tested our relationship, so we headed to the Napa Valley for some rest and relaxation. We stopped in Sonoma to visit the historic Gundlach Bundschu Winery, one of the oldest in California. Clara enjoyed the visit but didn't share my enthusiasm for hearing every detail of the winemaking process. I was blown away by the beauty of the vineyards and the appeal of its wines.

When I returned to Seattle, I found a case of its wine on sale at a local supermarket. I'd never purchased a case of wine. It seemed a reckless extravagance. As a grad student, I couldn't afford it, but I bought it anyway. I told myself I was exercising restraint by limiting myself to a bottle a month over the next year, noticing the subtle changes in flavor as it aged, but the credit card bill was scary.

I wanted to start making wine, but I couldn't afford to do it. Instead, my grad school friend Tom Remmers taught me how to make beer. My first batch started out fine, with a brilliant coppery color and a yeasty smell, but after a few days, it began to smell bad. I'd obviously done something wrong, probably contaminating the wort with bacteria. But I kept at it till I became adept at making India pale ales.

One fall, my Italian soccer buddies invited me to join them in making wine. They made it as part of a large group, pitching in to buy fruit from California. I joined the others as we stomped and crushed the grapes, making a party out of it. They added no sulfite to kill the wild yeast and no cultured yeast for the ferment.

Several guys worked in the restaurant industry. They brought eggplant parmesan, pasta with red sauce, veal scaloppini. We ate and drank during the crushing and gabbed about food, wine, and soccer. There was plenty of drinking, but no one got drunk. We were there for the whole experience.

Later, I took home a few bottles. It was bold, fruity, and rustic, and I was fascinated to see the whole process from grapes to finished wine. It was a lot more complicated than making beer, and the key ingredient, grapes, seemed endlessly complex. I gave up beer making and became a winemaker.

Tom and I made our first batch the following year, buying fruit with a group of other amateurs. Our Cabernet came from the Sagemoor Vineyard, one of the oldest in Washington. The clusters looked like gleaming purple jewels, bursting with ripe flavors. We rented a stemmer/crusher to process the fruit. As much as I enjoyed making wine with my Italian friends, I wanted to make a more refined wine. I added sulfites to kill the wild yeast and cultured yeast to the must. I transferred the fermented 1995 Cabernet into a thirty-gallon barrel in my garage. Gradually, it evaporated. I had failed to keep topping it off, adding other wine to ensure there was no air in the barrel. It seemed like a minor error, but it had major consequences. The wine oxidized, smelling and tasting like vinegar. All that beauty down the drain.

I was disappointed but not deterred. It was a technical issue, and we could fix it. I took a break from reading Hemingway and Faulkner to study books about making wine, to learn all the steps of the process, to buy the right equipment, and to pursue the best grapes. Over several vintages, we refined our process, making better and better wine and winning awards at the local fair, including a first for a white and a first for red wines. The quest was on.

As I sought to improve our wine, I consulted with local winemakers. They were very helpful, sharing their deep knowledge and experience. I justified these discussions as research for my food and wine stories, but my quest was becoming more intense. I was becoming obsessed. I was thinking about soil, grapes, fermenting techniques—all the time.

The Washington wine industry was expanding rapidly, and eventually I was able to get fruit from marvelous vineyards, including Ciel du Cheval, one of the best sites in the state. Was it time to take the next step and become a professional, or was that a hopeless fantasy? That's what I wanted to find out.

16

On Hallowed Ground

Visiting Bordeaux and the Great Growths

To understand how to make world-class wine, you have to taste world-class wine. For five years, I made regular pilgrimages to Bordeaux and Saint-Emilion, home of the five first-growth (premier cru) French wineries—Château Lafite Rothschild, Château Latour, Château Margaux, Château Haut-Brion, and Château Mouton Rothschild—among other outstanding houses. For anyone who aspires to be a true wine connoisseur, an understanding and appreciation of Bordeaux is mandatory.

For winemakers, knowledge and appreciation of these wines are essential. Most of the winemakers I knew held regular tastings of these wines if they could afford them. New-world winemakers wanted to understand the best of the old world, even if their vineyards and methods were different. Bob Betz at Betz Family Winery, Chris Upchurch at DeLille Cellars, and Michael Silacci at Opus One all tasted and appreciated the greatness of Bordeaux wines.

But it was not easy or affordable to taste these wines. You had to be an industry insider or wealthy to try them and gain access to the wineries. As someone without the bucks to purchase wines costing $1,000 and more for a bottle, I had to find another way.

On our tenth wedding anniversary, my wife, Lisa, and I planned a trip to Europe, including a side trip to Bordeaux. After landing in Paris, we flew to Bordeaux, rented a car, and visited Château Magnol and Barton & Guestier, the largest wine distributers in France. With their help, I managed to secure highly sought-after visits to Château Margaux, Château Mouton Rothschild, and Château Angelus in Saint-Emilion.

Château Margaux, one of the five first-growth Bordeaux producers and maker of some of the world's most elegant wine, lived up to the stereotype.

The mansion was an architectural jewel box of neoclassical design, framed by carefully pruned plane trees. The rectilinear lines of boxwood hedges surrounded the winery, making it appear more of a French garden than a vineyard. But just outside the garden, the carefully pruned vines took off in all directions. The view was the complete fantasy.

At the reception area, a friendly, gray-haired woman wearing blue dress pants, a white-striped shirt, and dark vest greeted us and showed us around the winery. The vaulted barrel room was cool and dark, with a deep, earthy smell. We passed workers racking wine, similar to what I did in my garage but with much more expensive liquid.

At the end of the tour, the woman poured Lisa and me a glass. The wine was dark, rich, dazzling, light on its feet though ten years old. The woman mentioned the wine could be aged for decades but was fine to drink now. I agreed and accepted her offer of another taste. I sipped and savored, thrilled to be sampling some of the finest and most expensive liquid in the world.

Afterward we returned to our hotel in Saint-Emilion, a fairy-tale town not far from Bordeaux. Unlike the more industrial Bordeaux, Saint-Emilion is small, with some two thousand residents, a walkable downtown, limestone caves stacked with wine, and a charming cathedral swirling with swallows. It's an especially lovely corner of France, with perhaps the highest concentration of great wineries, restaurants, and sheer aesthetic appeal of any wine area in the world. While there I met several Americans who gushed about the place. I loved it, too, and decided that it would be an ideal place to run a travel writing class, which would allow me to return regularly and drink in all the richness and history of Bordeaux. After talking with Lisa, I decided to do it.

The logistics were formidable. I had to book hotels, restaurants, and winery visits, as well as a tour bus and guides. Barton & Guestier helped with some of this. I did the rest. After all, it would allow me to continue my education in the region and its wines.

The class filled quickly, with fifteen students from all over the world. We met in the lobby of the Auberge de la Commanderie, an ancient stone building with stained glass windows and a grotto-like classroom excavated from limestone bed. There I welcomed my international students.

Death by Food and Wine

After touring Saint-Emilion on the first two days, we embarked on a bus ride to Bordeaux. While I was sweating the schedule (things always seemed to run late), the rest of the group delighted in getting to know one another and sharing stories of other trips, family photos, and published articles. We were a self-selected group focused on food, wine, travel, and writing. It was exciting and nerve-racking. I hoped all the visits would happen.

We arrived at the ivy-covered Château Magnol and sat down for our tasting. Conversation flowed with the wine and sumptuous meal of roast chicken. By the time we finished, we were running late, so I had to corral the stragglers and get back on the bus.

A downpour hit as we arrived at Château Lafite Rothschild, its tall turret sporting a French flag flapping in the downpour. Frédéric, our handsome and dark-haired guide, got the ladies' attention. We proceeded past ancient candelabras and dusty wine bottles, and entered a vast, circular concrete vault of the barrel room with gigantic pillars and a cool, musty atmosphere. After the tour, Frédéric poured each of us a splash of the velvety 1994 Lafite Rothschild, a classic Bordeaux smelling of cedar, leather, and tobacco. Student Erin Byrne described it as "the entire history of France contained in one sip."

By the end of the day, everyone was ready to eat. I booked reservations at Le Lion d'Or, a delightful restaurant with zinc countertops, cabinets full of first-growth wines, and the inimitable Monsieur Barbier. The bald, boisterous, mustachioed owner wore chef whites and carried around the carcass of a roasted pig, one of the region's signature dishes. I didn't have to worry about what to order.

M. Barbier shaved truffles on top of the pork roast, bringing a rich earthy taste to the dish. I ordered bottles of wine to go with it and the other dishes. Everyone dug into the dinner, talking, laughing, reveling in the company. By time I paid the check, everyone was stuffed and ready for a nap on the trip back to Saint-Emilion.

This trip was the first of five travel, food, and wine writing classes that I organized to Bordeaux. On each trip I visited at least one first-growth and several other wineries. I became adept at identifying the qualities of Bordeaux and Saint-Emilion wines: the earthy tobacco tastes of the reds, the tart lemony flavor of the whites, and the opulent, decadent flavors of the dessert wines such as those of Château d'Yquem.

It was a wonderful education and a chance for Lisa and me to get away. We didn't take our kids along until the fifth year, when I felt I was organized enough to add a new "wrinkle." I had no idea how challenging that wrinkle might be.

Les Enfants Terribles

It's a commonplace that the world is becoming smaller and that countries are collapsing into one big multicultural mélange with everyone drinking Coca-Cola, eating at McDonalds, and using teeth-whitening products. But the idea that we're living in one big happy new world is sheer fantasy, a concoction of commercial advertising companies. The vision of a new, borderless world order glosses over the stubbornness of national cultures, especially one as distinctive and durable as that of France.

Yes, the French are jogging now. Yes, they love to try out their English on you. Yes, you can see *The Simpsons* dubbed in French (a great and unremarked cultural breakthrough), but in fundamental ways the French remain stubbornly Gallic.

I discovered this the hard way on a recent trip. I'd visited France many times, but this trip was the first time Lisa and I brought our children: the twins, Danny and Nicky, eleven; and Marie, seven. We were staying for a week in Saint-Emilion, an enchanting medieval hill town outside of Bordeaux. I'd been there many times before and had brought in dozens of others as part of a travel writing class for The Writer's Workshop. You'd think this steady stream of business and publicity would have insulated me from the wrath of the owners should some unpleasantness occur. Alas, it was not so.

While I was teaching my class, my wife took the kids to the pool. The children dived for coins, shrieked with delight, and did cannonballs into the water. Nicky upped the ante, skipping coins across the water. He threw one so hard it struck a tile and broke it into four parts, which sank to the bottom.

Lisa was buried in a book, unaware of the infraction. "They were behaving like ordinary American children," she said. "They were having fun."

The hotel owners came out and berated Lisa about the broken tile and a long list of the children's offenses, chief among them being loud and excessively exuberant.

"There is a right way and a wrong way to raise children," said Madame, a well-coifed woman wearing a designer dress and high heels.

"It was an accident," Lisa replied. "We will pay for the tile."

"The swimming pool is for swimming only," Monsieur countered. "The children should not be so wild."

"If the children don't behave," Madame said, "we may have to ask you to leave the hotel."

Since the tirade was mostly in French, Lisa didn't catch all of the details, but the force of it astounded her. She felt like crying.

Later in the evening, I arrived back at the hotel as Lisa took the kids for a walk. "There's been an incident," she said, frowning. "I'll tell you about it when I get back."

At the front desk, the receptionist presented me with a plastic bag of "evidence" including the broken pool tiles. When Lisa returned, I got her side of the story. Yes, the tile was broken, but it was unintentional. "I don't feel welcome here," she said.

I considered leaving the hotel, but I was right in the middle of teaching my class.

"Look," I said. "It's too complicated to find another hotel. We need to stay here. We need to get through this. I'll talk to the owner tomorrow."

Lisa reluctantly agreed.

The next morning I went down to the reception desk to talk with Monsieur, a shy, formal man in his mid-fifties with dark, close-cropped hair and an earnest, agreeable demeanor. I have known him for five years, but our conversations mainly have concerned the price of the rooms.

"Bonjour, Monsieur," I said. "Je suis desolée." I apologized for the children's behavior and offered to pay for the broken tile.

"Monsieur O'Connell," he says, "there is a big problem with the children." He gestured wildly, indicating the enormity of the offense. "But the biggest problem is with your wife. It's one thing for the children to misbehave, they are children. They don't know. But it's another thing if a parent does nothing about it."

"I agree," I said, "but my wife doesn't know France well. She didn't realize this behavior would be a problem. Children are raised differently in the U.S. There is a *difference culturelle*."

Monsieur considered the phrase. He is not an unreasonable man. This phrase "difference culturelle" seemed to strike a chord with him.

"I have noticed that American parents are more permissive with their children," he said. "Perhaps this is a difference between the cultures, between a culture that has more liberty and openness and one that is more strict."

I nodded in agreement. "Exactement, Monsieur." I shook hands with him, appreciating his willingness to comprehend our point of view. "We will keep control of the children. They will only swim in the pool. No throwing of coins."

"Merci, Monsieur O'Connell," he says, shaking my hand. "Excusons nous," he says, finally pointing to himself. "Excuse us."

Crisis averted, I walked back to the room to tell the rest of the family. From now on, no skipping coins in the pool. No enfant terrible behavior. The children seemed to understand the gravity of the situation. We left Saint-Emilion without further incident.

But our trip wasn't over. How would we fare in places where I didn't know the hotel owners? Would total strangers take us to task for the boisterous behavior of our children? I knew how to behave as an adult, but I didn't know what was expected of children. There were no courses in French child-rearing for foreigners. We would just have to wing it.

We drove south, stopping at Carcasonne, a walled, medieval city with moats, stone walls, and slate-roofed turrets. I figured the kids would love it, but they were more interested in buying plastic turtles and eating ice cream than appreciating the architecture.

However, the boys' faces lit up when they saw a shop full of medieval weapons. There were swords, maces, chain mail, shields, spears. They spotted two large wooden swords. They had to buy them. I wavered. Who knew what international incidents they'd provoke? In the end, I caved. I wanted them to leave France with good memories. We barely managed to stuff the swords into our overpacked Renault.

We headed south to Collioure, a small port city surrounded by an ancient stone fortifications and groves of olive trees and with endless views of the wine-dark Mediterranean. The kids could swim, explore the old fort, and skip rocks in the sea without worrying about being out of line. That was the idea.

We booked rooms in the Hôtel des Templiers, a rambling, four-story hotel right off the waterfront. A former hangout of Matisse and Picasso, the walls were hung with canvasses by artists who had stayed there over the years, exchanging works of art for the hotel bill. There were no Matisses or Picassos but plenty of wonderful paintings.

The hotel reception was located in a bar on the ground floor, far from our fourth-floor rooms overlooking the harbor. When the children ran down the halls, they didn't seem to bother anyone, but we reprimanded them anyway, anxious to avoid repeating the experience in Saint-Emilion.

We ate at tourist restaurants at first. We kept it simple, the kids ordering steak and fries or chicken. We ordered duck confit or salade Niçoise draped with the tangy local sardines. They'd gobble their dinner and go outside to play on the beach while Lisa and I lingered over a bottle of chilled rosé.

I watched how the French children behaved at the restaurants. They ate with their parents and threw no tantrums. There was no pouting. No loud complaining. They were welcome as long as they behaved. If they did not, they were removed. There were no exceptions. The adults had less tolerance of misbehavior from kids than from dogs.

Over the next few days, we began to get the hang of things. We stuck to the noisy, informal cafés and ate breakfast on the hotel's terrace. Danny complained about having to eat croissants instead of cornflakes but took it no further. He sensed the rules were different here.

The kids enjoyed themselves, savoring trips to the ice cream stand, swimming in the sea, exploring the tide pools, and diving down to see the schools of fish. Outside they could yell and scream as much as they wanted; but inside the restaurants, they had to keep a lid on it.

The intense sun and relaxed attitude of southern France took hold; the tension in the hotel at Saint-Emilion was mostly forgotten. For our last night in Collioure, I booked a reservation at La Neptune, an elegant restaurant perched on a headland overlooking the harbor. I knew I was taking a chance, but I didn't want to eat in a cheap café.

We arrived early, hoping to avoid the dinner rush. The waiter, a slim, dark-haired blade of a man, moved skillfully and deftly around the outdoor deck. He smiled warmly and guided us to our table. There were no other children in the place, even though the restaurant had a children's menu. The other diners were older French couples.

He sat us at a table with a panoramic view of the harbor and the old fort. We ordered *steak frites* (steak and fries) for the kids, along with Coca-Cola, and *crème glacée* (ice cream) for dessert. Lisa and I looked at each other and hoped for the best.

The waiter brought us a bottle of rosé in a glass bucket of ice. The evening sun caught the bottle, turning it into a lovely shade of crimson. I began to let down my guard, relaxing into the moment.

Marie drew pictures of cats on a piece of paper. Danny read a book about submarines. Nicky paged through a comic book.

The first course arrived without incident. For me, local sardines with oil streaked so that they looked like a Matisse still life. For Lisa, a cool gazpacho soup. For the kids, small tenderloin steaks and fries.

"Is this a snack?" Nicky asked sarcastically but didn't push it.

The children ate their dinners quickly, digging into the potatoes fried in duck fat, a feature of French cooking they'd come to crave.

Then the main course appeared—tuna steaks—with a bottle of Collioure red. The steaks were perfect, charred on the outside, red on the inside. The wine was rich and robust and slightly cool—just the way I like it. The kids left the table to play in the tide pools below. There was no yelling. No shrieks of disapproval.

Lisa and I reminisced about the trip, joking about the incident in Saint-Emilion from the distance of a week. When the sundaes arrived, we called the children back to the table. To the waiter the children all said, "Merci." They quickly lost themselves in lapping up the ice cream. Lisa and I shared a café crème.

The older couples glanced wistfully at our table, perhaps remembering fondly their own years of raising children, the decades having erased the unpleasant aspects of the job. As we got up to leave, an older French couple beamed at us and beckoned us over.

"Elle apprend le Français," the husband said, pointing to Marie. She's learning French.

"Oui, Monsieur," I said, touching him on the shoulder with gratitude. We left the restaurant with lights shimmering on the harbor, the sun disappearing behind the Pyrenees, and the other couples talking and laughing over their dinners. As we walked back to our hotel, I reveled in the afterglow of it all: Enfants terribles no longer, our kids had finally become *les enfants français*.

This trip turned out to be my last to Saint-Emilion. Having visited all the first-growth and many other wineries, I felt that I'd accomplished my goals there. The incident at the hotel served as the frosting on the cake. Other wine regions beckoned, including Provence and Italy's Montalcino, with more knowledge to add to my wine making.

PART 5

California

17

Robert Mondavi

Transforming the Tastes of the Nation

When I visited the Robert Mondavi Winery in the spring of 1999, Mondavi was in fine form. On a sunny morning in April, we drove out to his beloved To Kalon Vineyard for an inspection. Phylloxera, a root louse, had devastated portions of the property but not yet infested the I block, a planting of prized fifty-one-year-old vines located in the heart of the Napa Valley.

Dyson DeMara, an assistant, stopped the green Chevy Suburban at the edge of the vineyard, and Mondavi hopped out. Wearing slacks, a blue dress shirt, and a sports jacket, the eighty-six-year-old strode briskly down the vine rows, his halo of white hair whipping in the wind as he scanned the gnarled vines.

Bright green leaves sprouted from the black trunks, indicating the plants remained immune to the pest. Mondavi spotted a bunch of grapes hanging from each shoot; the vines would need little thinning, pruning, and watering this year. Such subtleties would have been lost on most people but not on the winemaker, whose attention to detail was legendary. Having completed his inspection, he waited for DeMara to pop the cork on a bottle of 1996 Fumé Blanc from this vineyard.

"Look at it," he said, swirling the golden liquid in his glass. "It has depth, style, character." He sniffed it with his large beak of a nose, brought it to his lips, and washed the wine around in his mouth to bring out its full flavor. "It's as gentle as a baby's bottom but has the power of a Pavarotti."

I laughed at the mix of metaphors.

"Try it," he said, offering me a glass.

I took a sip, swirled it around in my mouth, and savored its rich flavor and bracing finish. "Delicious!" I said.

He smiled, and we got back in the truck. He dropped me off at the gleaming white mission-style building in Oakville on the St. Helena Highway.

I joined a tour group led by a vivacious young woman in a red dress. We filed into the interior of the winery past gleaming steel fermentation tanks, one of the hallmarks of Mondavi's wine-making style, emphasizing cleanliness and scientific experimentation over traditional techniques such as concrete and wood fermentation. Next, we entered the barrel room, with row upon row of sixty-gallon, purple-stained French oak barrels beneath vaulted ceilings—an underground cathedral of wine.

I felt like genuflecting, reveling in the smell of must, yeast, and wood. I marveled at the number of new French oak barrels, costing at least $1,000 apiece. The barrels imparted a sweet vanilla taste I didn't enjoy, but many people did, helping Mondavi create a huge market for his product.

Afterward, the guide led us on a tour of the grounds, where I took in an array of contemporary sculpture including the famous *St. Francis of Assisi* by San Francisco artist Beniamino Bufano. Saint Francis raises his arms protectively over a mosaic robe decorated with brightly colored birds roosting in a tree of life. The sculptures emphasize Mondavi's belief in the ties between food, wine, and the arts, and millions of people have found his vision of the good life appealing.

After seven decades in the wine business, Mondavi still loved his product and his work. While most of his contemporaries had retired or gone to the grave, the fit, trim Mondavi had no rearview mirror and only one speed—fast forward. But after building a multimillion-dollar business from scratch, putting the Napa Valley on the world wine map, and becoming America's most celebrated winemaker, what do you do for an encore?

Well, if you're Robert Mondavi, you aim higher. In addition to sponsoring Copia: The American Center for Wine, Food, and the Arts in Napa, he worked to refine the high-end winery Opus One. These efforts were in keeping with his desire to show how wine was an integral part of a civilized American life.

"Wine to me is passion," he says in his autobiography, *Harvests of Joy* (1999). "It's family and friends. It's warmth and generosity of spirit. Wine is art. It's culture. It's the essence of civilization and the Art of Living."

For Mondavi, this vision informed everything he did. He was a walking, talking advertisement for this way of life. Everywhere he went there was a spring in his step, a glint in his eye, and a passion in his voice as he proclaimed his gospel of the good life.

Mondavi's upbringing in a traditional Italian family profoundly influenced this gospel. Born in 1913 in Virginia, Minnesota, he learned early the connection between wine, food, and family life. His mother, Rosa, whipped up simple, delicious Italian dishes from olive oil, tomatoes, cheeses, pasta, and polenta. These meals served as the glue that bound the young family together.

The Mondavis supplemented these meals with strong, homemade wine made by his father, Cesare. Every autumn, he would bring home batches of wine grapes. Young Robert helped him stomp the grapes, press the wine, and store it for aging. Cesare made the wine as purely as possible with no extra additives. These early lessons in purity and simplicity became the foundation of Robert's approach to wine making.

Wine played a central part in their lives. They thought of it as a kind of liquid food, an essential part of everyday meals, not something to drink to excess. "I was brought up with wine," he says. "I never saw my family abuse it."

But not everyone considered the consumption of alcohol beneficial. The National Prohibition Act of 1919 banned the sale of liquor and beer, shutting down his father's saloon. Cesare suddenly needed a new career. After visiting Lodi, California, to buy grapes for the local Italian Club, Cesare decided to move his family out west to enter the produce and grape business.

At that time, Lodi was the grape capital of the United States. Mondavi grew up in the quiet farming community surrounded by other Italian families who shared the old-world values of love, discipline, and hard work, and the joys of wine and food. He and his brother, Peter, helped their father in the produce business. In turn, his father paid for their college tuitions.

An ambitious young man who strove to excel, Robert attended Stanford University, where he majored in business. He came away impressed with the school's academics but disgusted with many students' consumption of alcohol.

"The kids would have these beer busts," he says. "They'd drink beer, scotch, bourbon, and brandy. A third of them got plastered. I said, 'Are these people civilized?'"

The students' drunkenness had a powerful effect on Mondavi. He realized that not everyone shared his family's attitude toward alcohol. He understood that if he wanted to succeed in the wine business, then he would have to educate the American public about the benefits of fine wine and food.

After studying enology, the art and practice of making wine, at UC Berkeley, Mondavi took a job at Sunnyhill Winery in 1936. He didn't make much money, but he learned the intricacies of harvesting and crushing grapes and making and selling wine. And he loved the work. He had found his calling.

There was no fine wine business in California at that time. Wine was made in bulk, to be drunk quickly or stashed in a jug under the kitchen sink. Later on, it would be Mondavi's legacy to help create a market for high-end wines, but back then, he was content to buy grapes, make bulk wine, and ship it in tank cars to the East Coast for bottling, distribution, and selling. Even though the standards of wine making were low, Mondavi sought to excel. He set himself the goal of making the best tank car wine in America.

This goal did not satisfy him for long, however. When an opportunity arose to buy the historic Charles Krug Winery in Napa, he jumped at the chance. Since he did not have the money or credit to buy it himself, he prevailed on his father to purchase it. Cesare agreed, provided that Peter and Robert would work together in the business. The Mondavis' acquisition of that winery in 1943 was an essential step in Robert's dream of making great American wine.

"I started in the bulk wine business in the Napa Valley," he said. "From the bulk wine business, I began to taste the great wines of California: Inglenook, Beaulieu, Beringer Brothers, Wente. I said, 'Wait a minute, we can compete with them.'"

Mondavi's perfectionism eventually paid off. Charles Krug wines began winning awards at state fairs. The winery joined the top tier of California wineries. But Mondavi wasn't content. On a trip to Europe in 1962, he came away convinced that Krug's wines could compete with the best of Europe's, provided the winery made a commitment to excellence.

At the time, Mondavi didn't realize just how steep a price he would have to pay for excellence. For twenty-three years, he and his brother Peter ran Charles Krug together, building it into one of the most respected wineries in the Napa Valley. Theirs was a tight-knit family business, with their two sisters and parents also owning shares of the winery. The brothers clashed over the years, but with their father's intervention, they kept working together. However, Robert's ambition to make world-class wine strained their relationship to the breaking point. After their father died in 1959, the friction increased.

The conflict came to a head in November 1965. During a supposedly happy family gathering at Charles Krug, the brothers quarreled. According to Robert, Peter accused him of taking money from the winery. Robert denied it. He replied that he'd hit Peter if he said it again. When Peter repeated it, Robert hit him hard. The two men in their fifties fought like kids in the schoolyard. After it was all over, there were no apologies. A lawsuit followed. The fistfight and ensuing court battle divided the family and broke their mother's heart.

"We had a difference of opinion," Mondavi says. "I wanted to excel. I wanted the best equipment money could buy. My brother was much more conservative. He believed in going much slower and not spending all the money. My brother and I had honest differences of opinion. We did clash, and then I left the company. I think it's a blessing. I was able to excel, and he was able to do what he wanted."

The lawsuit tore the family apart, but it liberated Robert from Krug. The courts eventually ruled in his favor, giving him capital for his next adventure. With support from investors, Mondavi set out to create a new winery in Napa, the first since the late 1930s. His winery would have a radically different agenda—to make world-class wine. He called it the Robert Mondavi Winery.

Many in the Napa Valley thought Mondavi was delusional. There was no real market for fine American wine at that time. Americans preferred to drink beer, and if they did spring for a bottle of fine wine, it had to be French or Italian, not Californian, which was considered one step up from soda pop. Put the Napa Valley on the map? Change the eating and drinking tastes of the nation? What nonsense!

But the doubters underestimated Mondavi's drive. With his energy, optimism, and sheer tenacity, he steamrolled these objections as if they were toads on a road. Mondavi is not a large man, but he is a formidable presence with a hint of steel behind his sunny smile. He knew that the Napa and Sonoma Valleys had the potential to make world-class wines. All the ingredients were there; someone just had to take advantage of them. That someone was going to be him.

"You take the soil we have here, the terroir," he said. "It has its own style and character. I knew that we had the natural elements—the climate, soil, grape varieties—to produce wines equal to the finest, but our style would be somewhat different. I was convinced that we could do that."

Once his winery began producing premium wines, he believed that the market eventually would follow. To encourage this, he made education the centerpiece of his business plan. Through tours, tastings, and educational programs, he sought to introduce the public to his vision of the good life, one visitor at a time.

At that point, new wineries were not popping up like mushrooms in the Napa Valley nor were prices for fine wines soaring like tech stocks. Mondavi made progress through the late sixties and early seventies but hit a wall in the mid-seventies. The country's economic slide during the oil crisis took the wine market down with it. Mondavi's business slumped.

Then something extraordinary happened. An English wine merchant named Steven Spurrier organized a blind tasting of American and French wines in honor of the American Bicentennial of 1976. The event took place on the patio of the Hotel Intercontinental in Paris. The judges were all distinguished French connoisseurs representing the upper echelons of the Gallic wine establishment. After sniffing and sipping, the judges announced the shocking results: Three of the top four white wines were from the Napa Valley, including Mike Grgich's 1973 Chateau Montelena Chardonnay, which won first place. The first prize in reds went to Warren Winiarski's 1973 Cabernet Sauvignon from the valley's Stag's Leap Wine Cellars. The tasting thrust the Napa Valley into the world spotlight and created an instant demand for its wines.

Though Mondavi's wines were not chosen for the contest, the event vindicated everything he had claimed about the potential of California wine. Subsequent tastings only confirmed these results. In these tastings, Mondavi's wines garnered a large share of awards and prizes. Gradually his winery became one of the premier houses in the Napa Valley.

As the prestige of his winery grew, it attracted the attention of Baron Philippe de Rothschild, the owner of Château Mouton Rothschild, one of the most renowned wineries in the world. The baron wanted to form a partnership with a California operation, and he approached Mondavi about joining him. After a visit to Château Mouton in Bordeaux in 1978, Mondavi came away convinced the partnership would work. They would call the venture Opus One with the goal of making an American wine equal to the best Bordeaux. After spending millions to build a state-of-the art facility, they produced their first vintage in 1979. At its release in 1984, it sold out immediately at fifty dollars a bottle. Opus One became known as America's first ultra-premium winery.

Mondavi's business, along with the rest of the Napa and Sonoma wine industry, prospered as never before, but the future was by no means assured. A lingering neo prohibitionism dogged the business. In 1989 winemakers were forced to put warning labels on their bottles, as if wine were a kind of prescription drug. Mondavi sought to blunt the impact of the warning label by adding one of his own:

> Wine has been with us since the beginning of civilization. It is the temperate, civilized, romantic mealtime beverage recommended by the Bible. Wine has been praised for centuries by statesmen, philosophers, poets and scholars. Wine in moderation is an integral part of our culture, heritage, and gracious way of life.

This statement echoed his gospel of the good life and dovetailed with contemporaneous research on the health benefits of wine. Dr. Curtis Ellison, an epidemiologist who had worked at Harvard's School of Public Health, announced some startling findings regarding heart disease that were referred to as the "French paradox." The French eat rich, fatty foods and smoke and drink excessively, but they suffer fewer heart attacks than Americans. The cause of this discrepancy? Ellison speculated it was the red wine the French consumed. After he approached CBS's *60 Minutes*, journalist Morley Safer did a segment on the French paradox, concluding red wine was behind it.

Wine sales soared. Mondavi was pleased but not surprised. For years he had lauded the healthfulness of wine. Finally people were listening. The culture of wine and food was beginning to take root in American soil. To ensure it continued to grow, Mondavi threw his wealth and prestige behind the museum Copia in the Napa Valley. The facility opened in the fall of 2001.

In this way, Mondavi transformed the tastes of the nation, making wine and food a central part of American culture. He has taken the vision of Italian life inherited from his parents and grafted it onto contemporary life in these United States, helping to create a whole new market for fine wine and food in this country.

Why did he do it? Why not simply rest on his accomplishments in creating a billion-dollar wine empire? (In 2004 Constellation Brands acquired the winery for $1.36 billion in cash.) The answer lay with what he saw in wine. After pouring a glass of 1995 Robert Mondavi Reserve Cabernet in the winery's Cliff May room, Mondavi sipped the wine thoughtfully, reverently, savoring every nuance of its color, taste, smell, and texture. All of the pomp and circumstance surrounding the wine came down to this splash of liquid in a crystal glass, yet within that liquid lay a wondrous and measureless mystery.

"Wine holds up to our eyes light and color that no man could create," he says in *Harvests of Joy*. "In its essence, I believe that wine holds out to us all the order and wisdom of nature and of God himself, if only we have the patience and faith to pursue all the mysteries and truths waiting inside."

Before he died in 2008, Robert Mondavi cemented his reputation as an undisputed leader of the California wine industry, an innovative winemaker, and a shrewd, skillful, and energetic marketer. He was one of the most charming and forceful personalities I've ever encountered in the wine trade. He had a very good story to tell, and, by God, he was going to tell it.

After touring the winery, I sat down to lunch at the Cliff May room with other journalists visiting that day. Mondavi recounted the history of his winery with candor and humor, a version of what he'd written in *Harvests of Joy*:

> Wine has been with us almost since the dawn of civilization, and for centuries poets, painters, musicians, and philosophers have sung its praises. Even the Bible applauds its virtues. And wine to me is even more. When I pour a glass of truly fine wine, when I hold it up to the light and admire its color, when I raise it to my nose and savor its bouquet and essence, I know that wine is, above all else, a blessing, a gift of nature, a joy as pure and elemental as the soil and vines and sunshine from which it springs.

After he toasted the group, the winery hosted a fantastic lunch with different kinds of meats, cheeses, salads, and desserts—all accompanied by Mondavi wines, of course. These press events served an important role in his well-oiled marketing machine. While the event was certainly professional, it possessed a personality and warmth that I found appealing. Journalists from national wine magazines sat at the front of the room while journalists like me from more general interest magazines sat near the back. Mondavi made the rounds with everyone.

"How's the lunch?" he asked.

"Fabulous," I said.

"Great!" he said. He touched my shoulder, ready to move to the next guest.

"Could you do me a favor?" I asked. I brought out a bottle of my 1998 Les Copains Cabernet. "I also make wine," I said, perhaps impertinently. "Would you like to try a sip?"

"Sure," he said, sitting down. The waitress fetched a corkscrew and opened the bottle. She poured a sample in each of our glasses.

He swirled it around and then shoved his nose into the glass.

I know I was putting him on the spot. Would he be honest with me? Would he like the wine or spit it out? If he didn't like it, would he be too polite to say anything?

He took a sip, swished it around in his mouth. "It's good," he said.

"Anything I can do to improve?" I asked. I thought of Noël Coward's quip about criticism: "I love criticism as long as it's unqualified praise."

"Sometimes I'm not the best critic," he said, putting the glass down.

"Thank you for tasting it," I said. "I really appreciate it!"

He nodded and went on to visit other journalists. I went back to finishing my lunch, but I felt a flush of pleasure in my cheeks. He liked it!

18

Stag's Leap

The Search for an American Arcadia

On our honeymoon, my wife and I drive down to the Napa Valley. We stay at a cozy B and B in St. Helena, dine at fancy restaurants such as Terra, and rent bikes to tour the countryside. I try not to overdo the wine touring, as Lisa does not share this obsession, but I definitely want to visit Stag's Leap Wine Cellars. After all, a Stag's Leap Wine Cellars Cabernet won best red wine at the 1976 Judgment of Paris. The Americans were supposed to get trounced. Instead, American wineries like Stag's Leap Wine Cellars triumphed. It was the tipping point, the defining moment that propelled Stag's Leap and the California wine industry onto the world stage.

On a warm day in late June, we cycle along the Silverado Trail and stop at Stag's Leap, which sits at the foot of Parker Hill, a wooded knoll in the heart of the Napa Valley. Grapevines radiate in all directions, carpeting the rolling hills, soothing the soul with their ordered rows and chartreuse leaves. Nature and culture blend into one harmonious whole.

Parking the bikes under the towering oak trees outside the winery, we enter the tasting room. It is cool, dark, and stacked with oak barrels, and features a wide counter where a woman greets us and offers to pour us a splash.

Tasting through the whites and reds, I sniff and swirl. The wines are very different from sweeter, familiar names such as Mateus, Blue Nun, Black Tower. The pours taste more like the wines I tasted in Europe, with complex, alluring flavors such as blueberry and dark cherry mixed with spices like cinnamon. They have a tartness that makes my mouth pucker and an ineffable flavor that makes it hard to stop tasting. (I still have to bike back to the B and B!) The wine seems the perfect expression of this place. I vow to return.

In 2007 Warren Winiarski, the owner of Stag's Leap Wine Cellars, sold the property to Ste. Michelle Wine Estates and Marchesi Antinori for $185 million but kept his Arcadia Vineyard. Shortly thereafter, I return to Stag's Leap, hoping to rediscover the magic of my first visit.

"It's a beautiful place," says dark-haired, red-cheeked winemaker Nicki Pruss, who leads a tour of manicured gardens and Mission-style buildings. I hope this peek behind the curtain will not diminish my early impressions of the place. The air is thick with the yeasty smell of fermenting grapes. Workers bustle back and forth, bringing grapes to be crusher. I feel as if I'm walking on hallowed ground.

"Sometimes I just drive through the vineyards and think, 'This is my office,'" says Pruss. "I've truly been blessed."

I nod, thinking I'd feel the same way if I had the chance to work here. As we tour the facility, Pruss describes herself as chief "grape herder," coordinating the growing, picking, tasting, blending, and bottling. Her mantra of "restraint, balance, elegance" inspires the crafting of the winery's signature blends.

"Each block is like a color on a painter's palette," she says. "Each block is a slightly different expression of Cabernet. Some have bigger, more structural components. The soils, climate, and grapes—all make a difference in the blend. We are trying to become in tune with this place."

We walk outside the winery and look at the vineyards shimmering in the distance.

"I look at it as a stewardship," Pruss says of her job. "This is an important piece of ground to protect. I want to take it to the next level."

She demonstrates the Mistral destemming machine, which takes clusters of grapes and destems and sorts them, using a shaker table and an air jet to remove underripe grapes. "The goal is to make the grapes look like blueberries by the time they enter the fermenter," she says. "It allows you to select only the best berries. We're taking what we can from the past and moving it up one step higher. This allows us to go to the essence of the fruit."

After touring the winery, we enter the cellar to taste the lively whites and deep, rich Cabernets. Compared to the big tannic wines of Bordeaux, these wines seem soft and drinkable, the tannins hidden behind the gorgeous black fruit.

"It's an iron fist in a velvet glove," Pruss says proudly of the Cabernet.

The tasting culminates with the Cask 23 Cabernet Sauvignon, the winery's signature blend. The wine is dark, rich, and stylish but light on the tongue. It reveals the magic I found on my first visit, but it's concentrated, hinting at profundities only to be revealed in time.

"It's the top expression of our wine," she says, swirling it around in her glass before inhaling deeply. "It's a special place and a special product. We're taking what we can from the past and moving it up a step higher."

Near the end of the tour, we stop by the Hands of Time exhibit behind the winery, a wall with framed handprints of those who have worked at Stag's Leap and then moved on to found their own wineries or to work at some of the most famous ones in the Napa Valley, including Michael Silacci at Opus One. It's a who's who of the wine industry, a tribute to how Stag's Leap has transformed it.

Pruss came to Stag's Leap as an intern and is now the winemaker. The emphasis on tradition is clear. She hopes to build on the legacy of Warren Winiarski, who still lives on the property. I thank her for the tour. I feel there's so much more to discover, like peeling the layers of an onion.

Walking the Vineyard

On my next visit, I arrange a tour of the vineyards to see an entirely new dimension of the property. Passing through rows of carefully tended Cabernet vines, I follow vineyard manager Kirk Grace as he points out the unique attributes of the Fay Vineyard, the site that produced the 1973 bottle of Stag's Leap Wine Cellars Cabernet that won the 1976 Judgment of Paris.

"I'm charged with safeguarding the reputation of an iconic winery," says Grace, a tall, affable man, as we walk through rows of carefully tended vines. "Warren [Winiarski] was the visionary. He realized the greatness of the Fay Vineyard. We're carrying on the legacy."

Grace explains the sixty-five-acre vineyard includes steep, upper volcanic slopes and lower valley slopes containing clay. The vineyard's southern exposure ensures ripening, and cool night breezes maintain acidity in the grapes.

"The combination of the two sections of vineyard provides a nice palette for the winemaker," he says. "Mountain fruit and valley fruit combine for a great wine."

After the tour, we head back to the tasting room, which opened in fall 2014 and looks out over the fabled Fay and Stag's Leap Vineyards. "The new tasting room makes it easy to tell the story," Grace says as he uncorks the wines.

We try the perfumed Fay Estate Cabernet, the darker, more tannic Stag's Leap Vineyard Estate, and the rich, elegant Cask 23 Cabernet Sauvignon, which I remember fondly from my last visit. All of the wines exhibit a balance, depth, and proportion that defines the Stag's Leap style, a style pioneered by Winiarski and carried on by the new partnership of Piero Antinori and Ste. Michelle Wine Estates.

Stag's Leap depends on meticulous blending to achieve this style. After the tasting, winemaker Marcus Notaro leads me into the cellar to show how he accomplishes it. The air is thick with the smell of grapes and fermentation. We pass large stainless steel fermentation tanks containing wine from various parts of the vineyards. He stops at one, turns a small valve, and directs the wine into two glasses, handing one to me. It's wine from the Fay Vineyard I've just visited. "It's got a cherry cola note," he says. "The Fay has a lighter texture."

He tastes wine from the next tank from the Atlas Peak AVA of Napa. "This wine is more tannic," he says, swirling it around in his mouth.

I do the same, noticing the tartness of the tannins and trying to pick up other flavors from the yeasty young wine. I can discern the overall characteristics of the wine but not all the details he describes. Perhaps this ability will come with time.

He moves to another tank, takes a sample, and pours some into my glass. He swirls it around in his mouth. "This has potential for Cask 23," he says. "It's a season-long decision. Making the call on when to pick is one of the biggest things I do."

In blending Cask 23, Notaro and Grace follow in the footsteps of former owner Winiarski, seeking wines of balance and complexity. On my way out, I notice the winery has updated the back of the tasting room. The walls display paintings of the Ptolemaic and Copernican universes, as well as an orrery depicting the solar system. A Foucault pendulum swings back and forth in the barrel room. I've seen nothing like this in other wineries, some of which feature art and sculpture. This winery seems grounded in a rich cultural and historical context. Who built such a place? I have to find out.

Meeting the Wizard

On my next trip, I depart from a press itinerary to drive up to Warren Winiarski's house for an interview. I pass the winery and follow the road as it winds its way up Parker Hill. At the top, I find a modest house with a commanding view of Stag's Leap Vineyards, the Napa Valley, and the Mayacamas Mountains in the distance.

Warren meets me at the door and invites me in. A friendly if formal man, he looks like the college professor he once was, wearing khaki pants, an oxford cloth blue shirt, and round, studious-looking glasses. Now in his early nineties, he remains fit, trim, and passionate about wine.

He introduces me to his dark-haired wife, Barbara, and we enter the living room furnished with art books along the walls, a grand piano, and floor-to-ceiling windows looking out over the Napa Valley. The house is beautifully built and comfortably furnished, not a grandiose chateau-style McMansion. It exhibits the same balance and restraint as his wine.

He offers me a cup of coffee, and we sit down on a couch in the living room. He tells the story of how he got started in wine making. He began his career as a lecturer at the University of Chicago. As part of his graduate work, Winiarski, whose name means "son of a winemaker" in Polish, spent a year in Naples, Italy, where he fell in love with a life organized around food, wine, and culture.

When he returned to the United States, he began making wine, which fueled his desire to embark on a new career. He traveled to California to catch a glimpse of how he might do it. He spent a week working at the Martin Ray Winery in the Santa Cruz Mountains south of San Francisco. The winery seemed to embody his ideal of an American Arcadia.

"The week with Martin Ray was like living a dream. Here was everything: the bucolic location amid pines, communal dining, intelligent conversations. . . . There was almost a sacramental character to life at Martin Ray winery."

With a young family to support, Warren could not afford to go to school to learn wine making. Instead, he sought to apprentice at Martin Ray, learning by doing. Ray turned him down, writing that he seemed too independent to work at the winery.

Warren was disappointed, but the rejection did little to diminish his obsession with wine making. He sent out more letters to small wineries in the Napa Valley, looking for a job in the field. Finally he got a break. Lee Stewart at Chateau Souverain offered him a job. Stewart told him to arrive in the middle of summer to help with the harvest and crush.

In 1964 Warren loaded his white Chevrolet station wagon and a U-Haul trailer with clothes, furniture, and books. He, his wife, and their two children left his native Chicago bound for the Napa Valley, a modern pioneer family heading west.

When they arrived, Napa was a dusty agricultural valley with an informal, rural feel. There were no wine trains. No buses overflowing with tourists. No barn-size tasting rooms. No B and Bs. No clothing boutiques. No Michelin-starred restaurants. It was a kind of American Arcadia, rough around the edges, exactly what he was looking for.

When the Winiarskis arrived on August 1, they drove up to Howell Mountain to Souverain's winery. Stewart arranged for the family to rent a cabin below the winery on Crystal Springs Road. It had a wood-burning stove and a screened porch, with an orchard and a stream running through the property. The family could pick and preserve its own fruit and hear wind in the pine trees. Warren's dream was becoming a reality.

In the two years he worked at Souverain, Warren learned in detail about how to grow grapes and make wine. In his spare time, he experimented with making his own Zinfandel, using grapes from the Souverain vineyard. He talked, walked, and breathed wine making.

Though he hadn't planned to own his own winery, he took the plunge in 1965, using savings and loans to pay $15,000 for fifteen acres of land on Howell Mountain. At 1,900 feet, the vineyard was higher than most; Warren thought this elevation would inhibit the growth of the vigorous vines and make for a nicely balanced wine. Though he learned a lot from the venture, the site lacked adequate water. He sold the property in 1968 and went looking for another plot of land to buy.

In 1969 he visited Nathan Fay, a farmer who grew wine grapes for personal consumption. Like most home winemakers, Fay was eager to show off his wine. He offered Warren a taste of his 1968 Cabernet. Warren was blown away by it. He thought, *This is the best Cabernet I've ever tasted! It has a strong regional character but also a classical Cabernet character. It combines suppleness and strength.*

Warren told Fay that the wine was good but did not let on just how good he thought it was. When the Heid Ranch right next to Fay came up for sale, Warren made a plan to buy it, figuring the soil was similar. He continued to buy fruit from the Fay Vineyard and later purchased the property in 1986.

His academic training didn't help with the fundamentals of wine making, but it gave him a perspective that others lacked. His background in the classics helped him recognize the greatness of the Fay Vineyard. As a student of the classics, Warren valued the golden mean, referring to the relationship between the sides of the rectangle, as a model for beauty and balance. Though commonly applied to art and architecture, Warren applied it to wine—in particular, to the wines coming out of his Stag's Leap Vineyard (SLV) in the Napa Valley that exhibited an appealing balance and harmony.

"It's a combination of fire and water," he says of the SLV, speaking in a gravelly voice with an occasional dash of impish humor. "The fire comes from the volcanic soils of the upper part of the vineyard. The water comes from the lower part of the vineyard. The wines in the middle mediate between the two extremes, creating a harmonious whole."

This description explains the character of the Stag's Leap wines, which demonstrate this restraint and dramatic tension. They don't celebrate extraction and fruit for its own sake but fruit in balance with acidity. They take a much different approach than the big Napa fruit bombs, which garner gaudy wine scores but lack the subtlety and expressiveness of a Cask 23, for example, and don't go well with food.

To demonstrate how his wine pairs with food, one of Warren's ultimate tests, he invites me to dinner. At the Bistro Don Giovanni in Napa, the restaurant staff greets him warmly. He's a regular here, and we're quickly ushered to a table in the back.

"I'm part of the oversight of the vines at Arcadia, but I'm not doing winery marketing and financing," says Warren over a glass of Roederer Estate brut rosé. "It's returning to the fundamental part that makes for the sublimation of the fruit into something more meaningful for humans. I get a lot of satisfaction in being close to the vines and watching this annual cycle, watching the cycle closely and becoming sensitive to its mood. It's a bit like keeping in touch with the delicate changes of your spouse. You want to know what's happening."

Warren asks the waiter to decant a bottle of 2013 Stag's Leap Wine Cellars Cabernet Sauvignon, Napa Valley, a fortieth-anniversary special bottling of the version of the wine that won the Judgment of Paris. After the floor manager, Nazareno, decants the wine, Warren praises his skill, thanks him, and goes back to the menu.

"Don Giovanni is my favorite restaurant for the bistro style in Napa," he says. "It has a lot of Donna Scala [former owner, now deceased]. Her food was for the spirit. That goal is maintained by her husband, Giovanni, who still runs the restaurant."

Warren recommends the grilled octopus and beet and haricots verts salad for starters and the veal marsala as the main course, a great pairing with the SLV Cabernet. I go with his recommendation; I figure he knows what he's doing.

Though he has experienced many harvests, he hasn't lost his enthusiasm for them. "They're all different," he says. "I tended to be by the book earlier, but then I got more flexible, knowing that if I paid attention to what was going on in the vineyard there was more diversity. I was becoming more sensitive to what was going on in the vineyard and relying more on my taste because I began to see that this diversity could be a powerful assistance to making beautiful wine."

After the starters, Winiarski pours the SLV Cabernet to accompany the veal. "It's opened up," he says, swirling it around in his glass. "It's still youthful. It hasn't lost its baby fat. It's fleshy with a little edge that will soon be gone."

As with all great wines, the SLV Cabernet changes in the course of the meal, displaying new flavors and aromas as the dinner progresses. The wine complements the veal without overwhelming it, achieving a harmony and balance, making for a memorable meal.

"I like wines that have an overall composition that produces a sense of completeness," he says. "Such wines are not aiming at power as an end in itself."

Warren could be talking about himself and the world he's created at Stag's Leap.

"How do you achieve that?" I ask, hoping to discover the secret.

"You have to listen to the grapes," he says. "You have to go back to the vineyard."

That's exactly what we'll do tomorrow. I plan to meet him at the Arcadia Vineyard to learn how to taste the fruit to determine when to pick. We'll walk the vineyard, tasting the grapes to decide if they've reached perfect balance. Dinner ends with a peach sorbet and plans to meet tomorrow at Arcadia.

Visiting Arcadia Vineyard

It's easy to forget where wine comes from. When you buy a bottle at the store, it can seem just another product. Put grapes in at one end, ferment them, and pump out wine on the other end—simple.

But the process is much more mysterious and complicated, harnessing the power of nature to help make the best wine possible. This may sound easy in the abstract, but it becomes quite challenging in the vineyard, especially during the harvest, when unexpected rain or heat can change the grapes' maturation or destroy the delicate compositional elements of the fruit.

Nature has the last word in the vineyard. Nature doesn't always cooperate, as I was about to discover when I visited Warren's eighty-four-acre Arcadia Vineyard, which is located near Mount George, five miles east of downtown Napa.

Named for the Roman poet Virgil's idyllic land, Arcadia includes soils from an ancient inland lake containing the skeletal remains of diatoms. These soils give the Chardonnay from here a lively minerality and, like a good Chablis, an oyster shell taste that matches well with seafood. Winiarski bought the vineyard in 1996 partly because it provided fruit for Mike Grgich, the American winemaker whose 1973 Chateau Montelena Chardonnay won the best white wine in the Judgment of Paris.

I arrive at the vineyard on a sunny Thursday in September. Most of the white wine grapes have been picked. Now Warren and his team will decide when the red grapes are ready to harvest.

He drives up in a silver Lexus 2008 SUV and parks near a prefab mobile office in a grove of oak trees. Warren spends a lot of his time in the vineyard but again dresses as if he still could be teaching political philosophy, wearing a blue oxford cloth shirt and khaki pants. With a mane of gray hair worthy of a composer, he's not tall but cuts an impressive figure; his movements are precise and exacting, showing he knows what he's doing and why. He may not make wine anymore—the grapes from Arcadia will go to Stag's Leap Wine Cellars—but he's still on top of his game when it comes to the harvest.

"I bought the vineyard in part because Mike Grgich had made some great Chardonnay from here," he says of the property. "It was only later that I found out why."

We walk toward the office. I feel underdressed in hiking boots, light hiking pants, and a T-shirt, expecting to be tromping around the vineyard, but Warren doesn't seem to notice, which is fine with me. I'm here to learn some of the secrets that made him such a successful winemaker. We enter the office, which displays a map of the vineyard's blocks on the wall. Warren wants to discuss how the white grape picking went and strategize about picking the red grapes. According to him, the year was typical, with only some heat spikes near the end of the summer. The Chardonnay crop was moderate, with the heat near the end of the summer completing the ripening. "The quality was very good," he says, "due to the mainly moderate temperatures, giving slow ripening opportunities for subtle compositional elements in the fruit."

He introduces me to viticulture consultant Garrett Buckland, a strapping man in a plaid shirt and jeans who helps plan the vineyard operations. Coordination with his crew is key to having the harvest run smoothly.

"You did a marvelous job," Warren says to Buckland. "The fruit is good and uniform. It's really beautiful. It has expressive flavor with high minerality."

Buckland smiles, acknowledging the compliment.

Warren continues, reviewing the details of the white grape harvest. It turns out the crop was larger than expected. "The estimates and actual yield of the Chardonnay were quite different," Warren says.

"Crop estimation is very difficult to get right," Buckland admits, shifting in his chair.

Warren explains the crop is estimated at what's called hard seed stage, when the berry is about half the size it will be at harvest. After the young grapes are measured, a formula is applied to estimate the eventual crop. It doesn't take more than a few grams of difference in each berry to radically change the estimate. If a cluster is eighty to one hundred berries, just small changes in the berry estimate will greatly change the total weight. It's one of the many variables that make wine making challenging and unpredictable.

"The crop estimate is important because the winery needs to know the crop size to know where all the grape juice is going and which tanks it will go to," he says. "Every winery has a finite number of tanks, and they have to plan the use of the tanks. It's like a hotel. How many people are in your party? If you have four hundred, our hotel might not be big enough."

Having reviewed the white grape harvest, they move to the reds, which will be picked soon but not until they've reached the point of balance or dynamic tension that Winiarski prizes. The grapes will be sold to Stag's Leap Wine Cellars, but they have to pass muster with him first.

"From my point of view, I'm happy with every block this year," Buckland says. "We're in great shape. The fruit held up well."

"I agree in general," Winiarski says, "but from my point of view, Block 1 looks like it needs water." He noticed the grape leaves were turning away from sun. The vine does this to preserve its moisture, signaling it's experiencing too much stress. "We don't want to shut down the vines," he says. "We want continuous ripening toward maturity. The shutdown interrupts that continuity."

Winiarski explains that at Stag's Leap Vineyard he would notice young vines expanding as they grew. At a certain point, they would begin to contract and turn away from the sun.

"As I told André Tchelistcheff [legendary winemaker and viticulturalist] one time," he says, "I saw the vines we'd just planted getting bigger. Then one day they started shrinking, a sign that they were under stress. André agreed and himself applied the word 'shrinking.'"

Heat spells not only stress the vines, leading to evaporation of the fluid in the berries, but also increase the relative amount of sugar in the grapes. That leads to more extracted, higher-alcohol wines. which are definitely not Winiarski's style.

"How do you want to measure the evaporation?" Buckland says.

"Partly by eyesight," Winiarski says. He explains that at a certain point of ripeness, the skins get parched, showing signs of dehydration. "It looks like smile lines at the corner of your mouth," he says. "I see that and know the berry needs water. The softness of the berry is also a factor."

"According to weather forecasts, there are no major heat spells coming up," Buckland says.

Winiarski nods and thanks him for the update. With the meeting over, we go outside to taste the fruit in the vineyard and decide when the grapes will be ready to pick. We pass through long rows of Cabernet and Merlot vines along the rolling hills of the vineyard. There is no traffic noise, no tour buses; it truly seems Arcadian. He leads the way down the rows, plucking grapes, biting into them, and analyzing their flavors.

"It's a little tart," he says of one row. "But it has transparent flavors." He spits out some of the pulp into his palm and holds up one of the seeds. "See the color of the seeds? They're mainly brown. That means the seed is ready to do its job. Its job is to achieve immortality for the plant. The fruit is an attractant; it attracts in order to spread the life of the vine from one place to another. The seed is that part of the vine that longs for eternity."

We keep walking and tasting. It's important to get an overall idea of what's happening in the vineyard. Though the vines may look the same, they can vary widely from one part to the next.

"Berries at the top of the cluster are riper," he says, tasting and chewing on the skins. "There's a richness of flavor. There's a freshness in the fruit. The balance is there. I think it will be next week that we pick."

As always, he's looking for the point of harmony, the balance at "high noon"—the point at which before or after won't be as good. If picked too soon, the fruit could taste vegetal; picked too late, it could be prune-like and overly alcoholic.

"There's a richness and a lingering flavor," he says, tasting another berry. "Feel that." He hands me one of the berries. "It's not tight. It's soft and firm together. Take your index and middle finger, and press your skin near your cheekbone. That's close to what you feel when the berry is ripe."

He keeps walking, picking grapes, putting them into his mouth, chewing them, and then spitting them out. "Marcus is looking for signs in the fruit just as I am. He doesn't rely simply on numbers."

Winiarski notes that every winemaker tastes differently. Every winemaker has his or her own taste profile. "Marcus might want to put it into a blend," he says. "He's probably putting all of that into his overall equation."

We walk toward a barn next to the office. Warren opens up the garage doors to reveal a Kubota utility vehicle, which looks like a souped-up golf cart. We get in and set off up the hill to survey the vineyard as a whole. "Part of the reason I like this job is because I get to drive this vehicle," he says, grinning. When he heads up one of the steeper hills on the property, I have to hold on to the dash to keep from bouncing out. He points out that the vineyard has many different soil types, exposures to the sun, and other variables that a winemaker needs to take into account.

"Do you miss running the whole show at Stag's Leap?" I ask him.

"Yes and no," he says. "This more limited role gives me the freedom to return to some academic pursuits." He regularly teaches a summer classics course at his alma mater, St. John's College. "It's one intensive reading and discussion of a single work, recently *King Lear*," he says. "I enjoy that. I've done it for fifteen or sixteen years."

He continues along the perimeter of the vineyard. The view opens up over the Napa Valley, with the green rows of vines below. He stops at a section that is partially shaded in the afternoon.

"The attentive winemaker would see this section of the vineyard as more tannicky," he says. "Balance in the vineyard implies you have different elements and variation in the fruit. Then you work to harmonize the elements. You want to find something between hard and soft."

He takes my notebook to make a drawing of the golden rectangle.

"When making wine, you're looking for those elements," he says, pointing out the relationship between the sides. He's looking for the sensory equivalent of the golden mean, where there's a balance of opposites, a dynamic tension. "I'm not aiming at power as an end in itself," he says. "I seek overall harmonic composition in wine."

Just as he did for the SLV before he sold it, Winiarski applies this to the grapes and the wine coming out of Arcadia Vineyard's distinctive blocks, speculating about how the various blocks might blend together to create a harmonious wine. As we drive back to the barn, Winiarski notices yellowing vine leaves on some vines in Chardonnay Block 2, revealing a magnesium deficiency. "For next year, we need to provide an amount of magnesium to maintain vine health," he says.

As always, he keeps a close eye on the vines. This attention to detail is necessary to make the wines of balance that he loves. It's the care in the vineyard, not the latest technological gizmos, that makes the difference for him.

"In Napa, at a certain point, wine-making technology had been amplified," he says. "Many people were so impressed by the new tools that they devoted themselves to exploiting all of this power to improve wine: new centrifuges, pumps, crushers, specialized yeast, etc.

"Then, they realized they couldn't get the fullest expression of the grapes through this technology alone. They couldn't extract the beauty of the grapes without going back to the vineyard. This happened gradually in the mid-seventies and eighties. For example, the conventional eight-by-twelve spacing made things easier for tractors to move down the rows but did not improve the quality of the fruit.

"And then the question became how to grow better grapes? How do we improve the grape?" he asks. He keeps driving, answering the question with his actions, focusing his care and attention on the vineyard.

I mention that I'll be heading next to Opus One to learn from winemaker Michael Silacci, who once worked for Winiarski at Stag's Leap.

"Michael Silacci was one of the best winemakers who ever worked for me, maybe the best," says Winiarski. "He has a sense of what the grapes were saying to him. He always listened to the fruit."

On the way back to the barn, we pass the new Block 9, which is devoid of vines and ready for planting. Warren is planting this hillside block with a Mount Eden Cabernet clone from Santa Clara County. "I love that clone," he says. "Martin Ray introduced me to it back in 1963."

After spending a week at the Martin Ray Winery at the start of his career in 1963, Winiarski came away determined to become a winemaker. Thus, planting the Mount Eden Cabernet clone here allows his career to come full circle.

Rather than retiring, he's continuing to work into his nineties, making sure the vines have the right care, nutrition, and irrigation. After we say our goodbyes, I head back to Napa, relishing how much I've learned. I'm amazed how much time and attention goes into keeping a vineyard healthy and producing beautiful fruit. I'm excited to be learning how to taste grapes so as to pick them at their peak of flavor. I doubt anyone would pay me to do it, but at least I'm forming a clearer idea of how to do it.

Looking at his Chardonnay block after the harvest, he says, "The vines in this block look as though they have said what they had to say for this cycle of the sun. They're now on their knees, wanting to rest. I'm here thinking about how to help them prepare for their next awakening."

I'd love to stay and participate in the picking, but I have an appointment with Michael Silacci at Opus One to learn the intricacies of his harvest.

While I drive back toward Napa, Winiarski continues to watch over his vineyard. It's his attention to detail at the smallest level that allows his vineyards to produce such superlative grapes.

19

Opus One

The Place Made the Wine

It's 5:30 a.m. and the sky is dark and filled with stars. I punch the gate code at Opus One. Engines whir and the designer metal gate creaks open, allowing me to enter one of the most prestigious wineries in Napa. I pass rows of carefully pruned olive trees and manicured lawns as I approach the entrance. The building rises out of the vineyard, revealing a mix of classical European and contemporary California styles.

Olive trees and limestone colonnades frame the big wooden door at the official entrance. The building is a combination of European and American designs, with California redwood and stainless steel blended with cream-colored Texas limestone that looks as though it could have been quarried in France. The winery has a futuristic look, with some comparing it to a Martian spaceship having landed in the middle of a vineyard.

The facade is elegant, refined, and classical, just like the wine made here, but I won't be entering by the front door. I'm here to work and learn about wine making from the ground up. I find the back parking lot behind a large cinder block wall, its industrial form and function replacing the refined facade. This back end of the winery is where the physical work takes place.

I park my rental car and watch workers arriving. Today I hope to have a chance to watch the picking, to learn when to pick, and to tour the inner sanctum of the winery. Though I've tasted the sumptuous wine before, I've never had a chance to visit where it is made.

Opus boasts a distinguished pedigree. It opened in 1991 as a joint venture between Baron Philippe de Rothschild of Château Mouton Rothschild and Robert Mondavi of Robert Mondavi Winery, two of the most famous winemakers in the world.

Mouton Rothschild is a first-growth winery in Bordeaux, France, one of only five earning that high distinction. Having visited Château Mouton Rothschild twice, I came away impressed with the wine's elegance and refinement; it had a much different style than some of the big Napa Cabernets that have hogged the spotlight over the last decade.

As noted previously, Baron Philippe de Rothschild consulted Mondavi about forming a partnership, and after visiting the château in Bordeaux, Mondavi agreed. In 1978 they began their venture, Opus One.

In 2001 they hired Michael Silacci away from Stag's Leap Wine Cellars. With degrees in viticulture and oenology from UC Davis and the University of Bordeaux, Silacci had an ideal background for the job, bringing new-world and old-world perspectives to the position. Indeed, one of his classmates from Bordeaux, Eric Tourbier, serves as Mouton's current winemaker, and in his previous job at Stag's Leap Wine Cellars, Silacci followed owner Warren Winiarski's lead in crafting wines of balance, elegance, and restraint.

After visiting Winiarski's Arcadia Vineyards, I'm looking forward to seeing how Opus operates. How do they manage their vineyards? When do they choose to pick their grapes? How do they vinify the wines? How do they incorporate French techniques in their wine making?

Scanning the parking lot, I spot Michael Silacci. I walk over and introduce myself. He's a genial man wearing hiking boots, blue jeans, and a pile vest emblazoned with Tonnellerie Berger & Fils, a barrel maker. Despite the early hour, he bubbles with enthusiasm for the task ahead, chatting and joking with the workers. Despite his long experience making wine in California and in France, he hasn't lost his zest for it.

As a young man searching for what to do with his life, he previously traveled the world and ended up in France to pick grapes. "I was in Nantes," he says. "I fell in love with the experience of wine making and two-hour lunches." Before going to France, he had wanted to be a kindergarten teacher, but participating in the harvest changed the course of his life. He decided to become a winemaker.

After Silacci rounds up his assistants, we pile into his silver Land Rover and drive to the southwest block of the To Kalon Vineyard. The moon is out over Napa. The air is cool and still. Temperatures are in the fifties. A strong, sweet smell of grapes fills the air. It's easy to fall in love with the experience.

We get out and approach the rows where workers are picking Cabernet. Wearing orange safety vests and headlamps, they hand harvest every cluster, using a harvest knife with a curved blade to remove them, and ensure they arrive at the winery in optimal shape. This practice is in keeping with Robert Mondavi's initial intent with Opus.

"Long before we entered into the Opus One partnership, we knew that the more gently we handle the grapes, from the vineyard through the cellar, the gentler and more flavorful the wine is. For Opus One, gentle handling was a primary goal," Michael says.

The workers drop the clusters into small plastic picking bins containing no more than thirty-five pounds. Each bin is labeled with the name of the block where the grapes were picked. The picking team moves quickly down the row, treating the fruit with care.

As we approach a large tractor moving parallel with the rows, Silacci indicates that I need to climb its short steel ladder to a platform to get a better look at the picking.

"We want to become part of the team, not an obstacle," says he puckishly.

I nod and climb the ladder to the steel platform at the back of the tractor. I'm eager to learn as much as I can about Opus's wine making, but I don't want to get in the way, a common problem in learning the fundamentals of any field. The platform offers a bird's-eye view of the picking as the tractor moves above and between the vines, its wheels straddling the rows. The bright LED lights illuminate the grapes, making it possible for the workers to pick in the dark and out of the heat.

"The vines rehydrate at night," he says. "There's not much water in the soil. We've been having a warm spell. It's currently fifty-four degrees in Oakville."

Silacci jokes with Chris Tomkiewicz, the tall, serious assistant viticulturalist. There's an easy camaraderie among the Opus team, whose members seem to enjoy their work, especially during the harvest. This is the culmination of a year's effort, with Silacci serving as the conductor of the symphony, the maestro directing and inspiring the team to bring everything together for the finale.

The crew blurs through the vines. They strip leaves from the lower part to expose the clusters, which are plucked and dropped into the plastic containers. All of this takes place amid the rumble of the tractor's engine and the harsh bright light of the LED lamps. After a box is filled, a foreman puts it underneath the tractor. It is labeled and transported to the winery, where the grapes will be crushed. Everything is tracked and handled to make the best possible wine.

"Would you like an espresso?" asks Tomkiewicz.

"Thank you," I say, taking the small paper cup of the dark, steaming liquid. I turn to see a small espresso machine attached to the back of the tractor; it's a nice amenity when you're getting up early. The strong coffee cuts through the fog of my tiredness. I sip my espresso and watch the crew, marveling at their efficiency, wondering if I could work so skillfully.

Silacci points out the renowned Harlan Vineyards in the hills above us, Robert Mondavi Vineyards to the north, and Far Niente Winery to the south. It's valuable property, well drained and flat, similar to the topography in Bordeaux. The sky turns bright lemon in the east as the sun begins to rise. It's a wonderful spectacle. Why don't I get up this early more often?

After the crew completes picking the row, we pile back into the Land Rover for a quick trip into nearby Yountville to grab pastries and lattes at the renowned Bouchon Bakery. My *pain au chocolat* is flaky, gooey, and delicious. Washed down with the robust coffee, it reminds me of being in France. I could get used to this lifestyle.

After breakfast, we return to the vineyards to see if the next block is ready for picking. Tomkiewicz shows me a sheet outlining the criteria for deciding when to harvest. He analyzes the acidity, seed color, pluck-ability, and flavors such as green or cooked. We head down the rows, plucking berries and popping them into our mouths.

"All of this will give you an idea of when the wine will be ready," he says. "We're looking for the peak of flavor."

Early on, the grapes will taste herbal, then like a cherry, and finally like a blackberry or cassis, the preferred taste at the peak. After that, the fruit becomes overripe, tasting of plums or raisins. The window for the peak is fairly narrow, adding pressure to the harvest.

I grab a grape and chew it, trying to match the taste in my mouth with the categories he's described. The pluck-ability is firm, the seed color is beginning to turn brown, the acidity is good, and the flavor lies between green and cooked. My palate is not refined enough yet to be able to detect the peak of the harvest.

We head down another row, tasting from left to right and right to left, zigzagging back and forth, taking berries from the top, middle, and bottom of the cluster. "It should be as random as possible," Tomkiewicz says.

Despite my lack of experience, I get a sense of how to describe the fruit. My mouth is purple from the grape skins and dry with tannins, but I'm thrilled to learn how to do this. As I continue to taste, I detect blackberry and cassis flavors. It seems the fruit is close to its peak, but the final decision is up to Silacci and the Opus team. They reach this decision by consensus.

"It's like when you're driving," Silacci says. "You have a blind spot. I have a great team so they help me to avoid blind spots."

The sun is up by the time we arrive back at Opus to see the grapes crushed and fermented. The care and attention to the fruit at each stage is impressive. When the grapes arrive from the vineyard, they are hand sorted at a shaker table, with workers removing any stray leaves and the occasional black widow spider. A conveyor belt sends the grapes up a ramp called the stairway to heaven because these delectable berries are close to making the cut for Opus. A machine stems and sorts the grapes by size before they are fed into the crusher.

Many properties pump their wine under pressure from the crusher to the fermentation tanks. Not Opus One.

"We built the cellar on two levels. At harvesttime, the grapes arrive on the upper level, where they are crushed. Then, thanks to what we call our gravity-flow design, the must streams gently and naturally down into the tanks below. . . . From the beginning of the wine-making process to the end, there is no pumping, no pressure, no bruising of the fruit. We can see and taste the resulting gentleness and flavor in every bottle of Opus One."

After crushing, the grapes flow into a stainless steel fermenting tank where their journey toward wine begins. The grapes are inoculated with a mixture of cultured and wild yeast to start fermentation.

"The goal is to get to 50 percent wild yeast in the fermentation," Silacci says. "That will add another layer of complexity to the wine."

Using wild yeast demonstrates Opus's commitment to innovation. Many wineries, such as my own, use only cultured yeast, which is reliable, predictable, and commercially available. By contrast, wild yeast comes from the grapes themselves and can be unpredictable even if it more accurately reflects the terroir, imparting the character of the vineyard to the wine.

"We're trying to make wine with yeast from our vineyard," says Gordon Walker, a bright, bearded, red-haired, self-described fungal fanatic. He helps run the winery's native yeast program. "It's part of the sense of place to inoculate the must with some of the wild yeast."

After visiting the winery, I get a chance to tour the vineyards with vineyard manager Juan Martinez, a calm, deliberate, mustachioed man with thirty-plus years of experience at Opus and other wineries. Over the years, he's learned to be a kind of grape whisperer, pruning, planting, suckering, thinning, spraying, and overseeing the cultivation of the plants. It's said that wine is made in the vineyard, so I'm intrigued to see how Opus does this.

"We start pruning in January," he says, pointing to one of the vines. "Bud break is from March to April. We do suckering in May. That's intense labor."

He walks the rows, checking in on his vines, making sure they look healthy. Some 60–70 percent of the shoots will be clipped so the vine can devote its energies to the grapes, not the leaves or other shoots. Martinez and his team continue caring for the vines into the summer and early fall until the grapes are ready to pick.

Opus favors dense planting for low yields. In this, they took a page from Mouton Rothschild, which plants vines much closer together than most American vineyards do. According to Patrick Léon of Mouton Rothschild, "Vines need competition. It makes the grapes push harder. This way, each vine produces fewer grapes, but the grape quality is far superior. Narrow spacing demands more vines per acre and much more expensive pruning as well. As a result, the cost of producing Opus One is much higher than the cost of producing many other fine wines."

I next visit with Tomkiewicz to get his perspective. "We have a mono crop here," he says, pointing to row upon row of vines. "We fight a constant war against pests, viruses, and fungi. For sustainability, we need biodiversity."

Driving around in a small cart, Tomkiewicz shows how cover crops replenish the soil, bringing back the insects and wasps to keep the vines healthy, thus increasing biodiversity. "We're trying to let the land speak through the grapes," he says. "We want a unique expression of a site."

After touring the vineyard, I head back to the "fishbowl," an enclosed windowed lab where we'll taste some recent vintages to see the results of the care and attention lavished on the fruit.

Silacci takes out wine glasses and covers for them. He uncorks a bottle of the 2010 Opus and pours it into the glasses. He places the cover over the glass and swirls the liquid around to release its aromas.

"What does this remind you of?" he says, taking off the cover.

I sniff the liquid.

"I'm getting dark fruits and spice," he says.

"Yes," I say. "I'm getting chocolate too."

We put the covers back on and swirl the glasses. When we sniff again, we get an even stronger whiff of the wine.

"Taste is smell," he says, taking a sip. "I'm getting some dark bitter chocolate with a long-lasting finish like church bells at the Vatican."

I laugh at the exactness of the image. I take a sip and swirl it around in my mouth. The wine fulfills the promise of the aroma—deep, rich, bursting with dark fruits, chocolate, crushed herbs, and a pleasing minerality in the finish that lingers like the tintinnabulations of church bells.

Next, he pours the 2012. We cover the glasses and swirl the wine. When he takes off the covers, we sniff. Dark fruit seems dominant in this vintage, with a similar aroma and taste to the 2010 but having its own distinctive mix of elements.

"Just one word," Silacci says. "Yummy!"

I nod and smell, swirl and sample, caught up in the magic unfolding in the glass. The wine reveals the care, work, planning, and, yes, poetry I've witnessed today. I compliment him on the wine.

"Thank you," he says. "But it's not about me or the team. It's about the place. It's about the vineyard."

Taking one last sip, I get ready to leave. This visit has been one of the most detailed, comprehensive ones I've ever experienced, including ones to the first-growth French wineries of Châteaux Mouton Rothschild, Lafite Rothschild, Margaux, Latour, Haut-Brion, and d'Yquem. There's an energy, vitality, and joie de vivre at Opus that reflects Silacci's leadership. When I mention this, he deflects the praise.

"Whenever I feel cocky," he says, swirling the wine around in the glass, "I remind myself that 80 percent of wine quality comes from the vineyard, and I think about the great 1945 Bordeaux, one of the greatest vintages."

Silacci believes winemakers need to respect and express the place. He sees the 1945 Bordeaux as a great example of this aspect of terroir. With the men at war, the young and old pitched in to pick the grapes and to stem and crush the fruit. With such minimal intervention, the wine might have been forgettable, but the results were spectacular.

As Silacci says, "The place made the wine."

20

Cain Vineyard and Winery

Letting the Land Speak

Cain Vineyard and Winery sits high in the hills above the Napa Valley. The 542-acre property contours around a vast, open, east-facing bowl, following the twists and turns of the curvaceous, idiosyncratic, and spectacular hills. It rises from 1,400 to 2,100 feet, revealing many distinct microclimates and soil profiles. It's also home to browsing deer, soaring hawks, and flocks of wild turkeys.

Located in the Spring Mountain District on the western edge of the Napa Valley, the vineyard straddles the border between Napa and Sonoma. It's about as far as you can get geographically from the many wineries such as Opus One located along Highway 29, the main wine-touring route through the Napa Valley. It's also a distance philosophically in its approach, emphasizing an intense focus on terroir in wine making. Cain makes three Cabernet blends, each from unique vineyard sources and with a distinctive approach: Cain Cuvée, a smooth bistro wine; Cain Concept, a silky, age-worthy wine sourced from other Napa vineyards; and Cain Five, the signature blend from the vineyard's best lots.

The steep, varied landscape of the Spring Mountain District makes for challenging terrain, with its thin, rocky shale and sandstone soils prone to erosion and the limited rain during the dry growing season. But these conditions hold out the possibility of making fabulous wines of character and subtlety as the grapes develop more intense flavors as they struggle for water, nutrients, and sun.

Winemakers here must walk a fine line, making sure their vines get enough water and nutrients but not too much. If they get too much, they put all of their energy into the foliage rather than the fruit. To make wines of character, the vines must be healthy but stressed so they put their energy into the grapes.

Fostering this process requires great attention to detail in this varied terrain, making large-scale production impossible.

Such terroir attracts strong-minded iconoclasts who often go their own way in wine making. Chris Howell of Cain Vineyard and Winery is such a person. He is passionate about the place, so devoted to the idea of terroir that he uses no cultured yeast in his wine; all of it is fermented using native yeasts. This is just one of ways he seeks to express the land through his wine, displaying perhaps the lightest touch of any winemaker in Napa.

A graduate of the University of Chicago in "ideas and methods" and of the University of Washington in chemical engineering, Howell is thoughtful and articulate with an understated intensity about him. He's refreshingly contrarian in his approach to wine making, always searching, probing, and experimenting to bring out the nuances of the vineyard.

I've written about him before regarding food, wine, and travel in Napa and Sonoma, but now I'm looking for a more in-depth story and doing research for my wine making. I'm not interested in wine scores or tasting notes but in the process by which they are achieved. I'm looking for a full immersion in the wine-making experience. I want to learn about it from the inside out. I'm hoping that understanding the process will lead to a fuller appreciation of the final product.

To do this, I set my alarm for 4:30 a.m. I need to drive from the Harvest Inn by Charlie Palmer in St. Helena and arrive at Cain by 5:30 a.m., not far as the crow flies but up an incredibly winding road. I shower, grab a cup of coffee, and hit the road. The Spring Mountain road twists and turns as it gains elevation, passing houses on the outskirts of St. Helena, climbing through oak and Douglas fir, horse farms, vineyards, and orchards. The rental car strains to navigate the hairpin turns. Even though I've been here before, it's hard to follow the road in the dark.

When I arrive at 5:30 a.m., I marvel at the profusion of stars: Orion's Belt, the gleaming W of Cassiopeia, and the familiar shape of the Big Dipper shining brightly above the Napa Valley.

Chris Howell is waiting for me. "Time to put your boots on," he says. "It's time to get to work."

I grab my notebook and pen. "Don't worry about that," Chris says. "We're working today."

I take it anyway. Will I get fired for insubordination on the first day? I hope not. I want to participate in the harvest as well as document it. After visiting Opus One, I'm looking forward to seeing how Cain Winery approaches the harvest.

I follow Chris through the darkness. We get into his white Ford Explorer and head down the steep, curving roads through the vineyards. Everything seems curved here in contrast to the rectilinear lines of Opus.

Chris parks at a barn where the workers have assembled to get ready for the pick. He hands me a red safety vest, a headlamp, gloves, and pruning clippers. I join the circle of workers ready for the early morning warm up. Ashley Anderson-Bennett, associate vineyard manager, leads everyone in stretching. No one is chanting om, but the goal is similar to that of a yoga class—limbering everyone up to prevent injury or soreness.

By the end of the stretching, I'm breaking a sweat. It's seventy degrees. Over the last two days, the temperature has spiked to one hundred degrees during the day, causing some of the berries to dry out. As noted previously, managing the weather is one of the challenges of harvest: Pick too early and the wine may taste "green," or vegetal; pick too late and the fruit could be dried out, leading to wines tasting "cooked," or like raisins.

After discussing where to pick, Chris and Ashley direct the crew. I hop up into the back of a pickup truck and head down the road. The other vineyard workers speak Spanish, only a little of which I understand.

One of the men hands me a gray picking bin, some thirty by eighteen inches wide. He and the others start working the rows, picking the Merlot grapes. A worker tackles one section while the others leapfrog ahead, carefully culling the clusters of grapes and dropping them into the bin.

I start picking, using the clippers to cut the clusters. I try to move quickly and efficiently, but I'm much slower than the others; figuring out whether I need to clip a cluster or simply pluck it takes a while. Judging the maturity of the fruit comes down to feel.

I fill up my first bin and hand it to the man driving the tractor with a tote behind it. He nods, dumps it into the tote, and hands it back to me. Then I move back down the row, walking ahead of the line of men and picking back toward them. My knees and neck ache from bending down to pick the grapes, but I keep going.

The men work quickly, joking and laughing, as they harvest the fruit. I try to keep up with them, disappearing into the work, testing the resistance of the grape stem for plucking or clipping, inhaling the fruity perfume of the grapes, and keeping the spiderwebs and straw out of my bin.

At the end of the row, I hand my bin over to be dumped again into the tote. I wipe the sweat out of my eyes, drink some water, and look out over the Napa Valley. The sun is starting to come up. Tiny pinpricks of lights in St. Helena glimmer below us in the distance. It's a moment to be savored, but I don't have time to savor it. The men take off their vests and headlamps, and start picking another row.

I join in, moving down the row, selecting a spot with lots of clusters. The men move past me, kicking the empty bins down the row, passing the filled bins over the vines to the waiting tractor.

At the end of the row, I stop for another cup of water. Abbey, one of the other workers, introduces himself.

"I'm Nick," I say in my limited Spanish.

"Rapido," he says, pointing to me.

"Muchas gracias!" I say, pleased for the recognition.

"¿De donde es usted?" he asks. Where are you from?

"Seattle."

"¿Hace frío?" Is it cold?

"No, es como San Francisco. ¿De donde es usted?"

"Napa," he says.

"¿Caro?" I ask.

"Mucho caro," he says. Very expensive.

We keep going until we finish the block. It's 8:30 a.m., the sun is up, and the temperature is rising. I grab another drink of water and stretch. My back and knees ache from the picking.

Chris drives up with Ashley and François Bugué, associate winemaker. Chris motions me over. He wants me to help taste the fruit to decide which block to pick next.

He leads the way to Block 10. We move through the rows, picking the grapes, tasting them, and analyzing the flavors, just as I did at Opus.

"This is a very sensitive block," says Chris. "It could collapse from the heat."

"There's no shriveling," says François, inspecting the fruit.

"We'll come here next," says Ashley.

"It's the right time," says Chris.

I taste along with them, following the process by which they make a decision to harvest. This is complicated at Cain as the steep hillsides and varied terrain make it hard to generalize about the vineyard as a whole. Every block must be tasted and analyzed before it is picked.

We move to the next block, winding our way around the huge open bowl of the spectacular Cain Vineyard. We keep picking and tasting, trying to determine when to harvest. This decision is also based on the varietal, whether Cabernet, Merlot, or other; the sun exposure; and the history of the vineyard—all critical knowledge that Chris has acquired over his many years at Cain. He started in 1990 and has worked here since. We pass through a block of vines planted in 1987; the first vines were planted in 1981.

"These vines were here before me," he says. "I love that idea."

We keep moving, picking, and tasting, caught up in the excitement of the harvest. The rows undulate into the next block.

"What do you think, François?"

"Pick, pick, pick," François says.

"Taste and smell," Chris advises me. "Every vintage is different."

Since this harvest is my first at Cain, I have no context for what I'm tasting and smelling. It tastes good, with deep plum flavors, moderate acidity, and a freshness that is appealing.

Hiking up a steep dirt road to the next block, I spot a black plastic drainpipe along the side. "What is that?" I ask.

"It's a holding tank to keep rain from eroding the vineyard," he says. "We get a lot of rain in the winter. Erosion control is essential at Cain. Soil is the most important thing. Keeping it intact and alive is the highest priority. It's a different world here than it is down in the valley."

It's midmorning now. After tasting through the blocks, Chris, François, and Ashley have a plan for what to pick next. I'm amazed at the variation in tastes of the fruit, depending on the hillside exposure, elevation, and canopy used to train the vines. Now I have a much better idea of what they're looking for in the fruit.

"If terroir is the goal, we want to perceive it and respond to it," says Chris, summing up the morning's tasting. "We want to make wine that tastes like it grew in a particular place."

We head back to the winery to check on the fermentations already underway. Cain is in the middle of its harvest, and Chris must juggle the tasting, picking, crushing, pressing, and fermentations. During the harvest, he works seven days a week for weeks at a time, but it's an exciting time, the culmination of a year of work,

"We are here to interpret the vineyard," he says, handing me a glass. "We aren't really that surprised at what the vineyard holds, but we never know what the weather holds."

There are many steps from the picked fruit to the finished wine. We taste the first of these steps right after the grapes are crushed and the fermentation begins. The liquid tastes like grape juice with just a hint of something spritzy on the tongue.

He swirls the juice in his glass and takes a sniff. "I don't want to make formulaic wine," he says. "I don't want the process of wine making to overwhelm the vineyard. Each wine has to have its own identity, derived primarily from the vineyard. Our job is to get to know the fruit and allow that expression to come through. We could pick too soon and not express the vineyard. Or we could pick too late, and it would taste like overripe fruit. The more we impose on the process, the less the individual vineyard will come through."

We pour the juice back into the tank and move to the next one. Again Chris fills our glasses with the fermenting liquid. This batch seems closer to wine, with a whiff of yeast and the tingle of what could be alcohol.

"Wine is about transcending fruit," he says. "It's like the difference between cheese and raw milk. You don't expect the cheese to taste like the milk. Wine is not like jam; it's a more complex, involved thing. Wine is about transformation, not preservation. There's something about wine that is magical."

We taste from a vat that has fermented for a couple weeks. It tastes less like fruit and more like wine, with interesting layers of flavor amid the fruity smells of the yeast.

"We're going against the uniformities of wine," he says, sniffing and tasting. "We want to enhance the diversity of wine. In the world of wine, diversity is a core value. At the Safeway in St. Helena, there are a thousand different bottles of wine. If wine were just a uniform manufactured product, we wouldn't be interested in all those."

Chris leads the way to the cellar to taste a 2015 Syrah in barrel. He pours a splash into our glasses. I swirl, sniff, and taste the wine. The grapes have changed remarkably since they were harvested but resemble the fruit I've tasted today, reflecting the powerful influence of the vineyard.

"Wine allows you to reconnect with nature," he says, swirling the wine in the glass. "It's hard for a lot of us to connect with nature. Many of us live in an urban environment. We turn on the light, but we don't know where the electricity comes from. I think people want to reconnect with the earth. Wine allows you to do that."

I thank Chris for the tour and hand back the glass. It's been a very instructive morning. I've witnessed and participated in some exciting new trends in wine making. I've learned about his approach from the ground up, including picking the grapes, evaluating them to determine when to pick, and observing a fermentation with wild yeast. His approach resembles that of Warren Winiarski and Michael Silacci but reveals more of the wild side of the vineyards with their unexpected smells and flavors.

A few years back, I joined Chris for dinner at Terra, a terrific restaurant in St. Helena. He brought several wines to taste during the meal, including a 2004 Cain Five, his top blend. The wine exhibited a strong umami, or meaty taste, that was more savory than sweet and smelled vaguely of barnyard. The smell and taste of the wine made for a perfect pairing with the cheese. I'd never had anything quite like it.

Howell excuses himself to check on one of the other blocks. I help François and Quentin, a young intern from near Bordeaux, France, clean out the wine press. They hose down the outside of the stainless steel press, being careful to remove all the grape skins and seeds. Then the inside needs to be cleaned.

I volunteer to climb inside with the hose, scouring the remaining seeds and skins from the press and getting myself thoroughly soaked. It's a relief to wash the sticky grape juice from my face, hands, and shirt. I hand the hose to François and an impromptu water fight breaks out, making sure everyone gets cleaned off—a kind of ritual baptism into the harvest at Cain.

PART 6

A Question of Balance

21

Taking Stock

Should I Go Pro?

Energized by my visit to California, I vow to be more scrupulous about every aspect of my own wine making, including racking, transferring the wine out of barrels, washing the barrels, and returning the wine to them. As wine ages, yeast, stems, and grape skins filter down to the bottom of the barrel. To clarify the wine, these solids need to be removed every few months. This process is called racking, a tedious but necessary task in making great wine. People consider wine making a romantic activity, but there's a lot of drudgery to it.

Before racking, I go downstairs where I store the four sixty-gallon barrels in the back of my garage. These beautiful French oak barrels—mahogany in color, stained purple with wine, showing a tight wood grain, carefully carved staves, all bound with steel hoops—are works of art in their own right. I keep two barrels on the bottom rack and two barrels on the upper rack. Like so much in wine making, the barrel rack is simple and efficient, the design honed over centuries.

Cement walls surround the barrels on three sides, keeping the area dark and cool most of the year. In the summer, I use an air-conditioning unit to prevent the temperature from getting above sixty-five degrees. If this were a professional winery, I'd have a dedicated facility with four walls and a consistent temperature of fifty-five degrees. As a home winemaker, I have a limit to the time and expenses I can spare.

"Can you move some of that wine stuff out of there?" Lisa says from upstairs. "It's filling up the garage."

Nodding, I stack some of the carboys in the back and push the wine supplies to a corner, trying to placate her.

Tom, Bruce, Theresa, Gregory, Chris, and Andrew will appear soon, ready to work. As much as I like to socialize, we need to stay on task. One year, we got distracted and put the wrong wine in the wrong barrel. We had to siphon it out, adding another hour to the job. Wine making may see like a fun, light-hearted activity, but it is exacting work.

I fill a glass wine jug with water and bleach, fire up the pump, and run it through a plastic hose, cleaning it. Then I run water through the hose to remove the bleach.

Tom arrives first, checking things out. I wash the glass jugs in the white plastic sink, stained purple from years of cleaning the barrels and bottles.

"I wish we could get a stainless steel sink," he laments.

"One of these days," I say. I don't need to put more money into this "hobby," which costs several thousand dollars a year and a lot of time and attention.

The goal tonight is to empty the wine from the barrels into plastic buckets, wash out the barrels, and then return the wine to the barrels. It sounds simple, but the more barrels the more complicated things become.

Tom uncorks a bottle of Syrah to compare with our own. Others bring bread, cheese, and cold cuts, adding to the festivities.

We put one end of the racking wand into the top of one barrel and the other end to the hose attached to the pump. The state-of-the-art wand is stainless steel with a curved *C* shape and valve at the end to prevent it from sucking up too much gunk. It's definitely a step up. Every year we try to improve.

The pump whines and hums as it sucks the liquid out of the barrel and into the plastic bucket. We dip plastic cups into the wine and taste it, noting its flavor and color.

"It's a rambunctious wine," I say, joking about its youth.

The others add their descriptors: fruity, savory, delicious, obstreperous, bad ass. Everyone helps themselves to bread, cheese, and meats as we catch up on what's happening with family and friends.

Amid the yakking, I hear the pump make a high whine. I turn it off, climb above the barrel, and shine my flashlight into the hole at the top to see only sediment left inside. Bruce and Andrew carefully lift the 120-pound barrel off the rack and carry it outside to wash it out.

One down, three to go!

Tom and I strategize about the order of racking. We do the next barrel on the upper rack. Then we tackle the lower barrels, making sure to rotate them. We use a mixture of neutral and new oak barrels. As much as we like the smoky, vanilla flavor a new barrel imparts, we don't want it to overwhelm the fruit.

We make two styles of Syrah. One is a free-run version, which we call our Bandol style after the wines of Domaine Tempier in Provence. We also make a pressed wine, leaner and more tannic, that we call our Côte Rotie style after the wines of the northern Rhône in France.

After the barrels and buckets are washed, Tom and the others go upstairs to share some food and wine. I'm left cleaning up in the basement, as the hoses and siphon still need to be washed. It's been a long evening. I tidy up a few more things and then join them, leaving the siphon in the last barrel. I'll finish tomorrow.

Upstairs we share a bottle of wine, usually one I've brought back from trips to Europe: Domaine Tempier or Domaine du Vieux Télégraphe from Provence, Bodegas Muga from Rioja, or Biondi-Santi Brunello from Montalcino in Tuscany. Tonight I uncork a bottle of Cain Five, a superb Napa blend made by Chris Howell. Great wine needs to be shared and tastes better in the context of food and conversation.

We sip and admire the wine, getting a sense of the high bar it sets, something we can aspire to. After everyone leaves, I take a shower and fall into bed exhausted.

When I go down to the garage the next morning, a huge lake of wine confronts me. It leaked out of the siphon I left in the barrel and onto the floor. I quickly remove the siphon, grab some towels, and start damage control. I chide myself for not finishing the job. Missteps in wine making can be unforgiving.

After cleaning up, I use the topping wine to fill the barrel. It comes up short. I don't want to leave any air in the barrel lest the wine oxidize and become vinegar. I drive to the store to buy some local Syrah to top the barrel.

After topping it, I can relax. I'm glad I'm responsible for only four barrels, not four thousand. As much as I enjoy doing this work on a small scale, I wonder if I'm cut out for this as a full-time gig.

22

Wine with Dinner

Making wine is a social activity. You can't do it all alone. When it's time to rack again, I invite one of Les Copains' members, Chris Olsen, over to help. As a bribe, I promise to make dinner and uncork one or our oldest wines, a 2000 Cabernet, that we made the year my daughter, Marie, was born.

One of the unique things about wine is the history it contains, not only the history of the grapes, but also the associations and memories attached to them. But will the 2000 be any good? Will it only have sentimental value? Will we take a taste and then dump the rest down the drain? I keep the garage cool, but I've had off bottles over the years.

Chris is taller than six feet and solidly built, so he's a good choice for racking, which involves lifting and carrying the empty sixty-gallon oak wine barrels. I've known him since high school, where we sat next to each other in algebra class. He went on to become an accountant while I became a writer.

After washing out the pump and siphon hose, I remove the bung from the first barrel and insert the siphon. When I flip the switch, the pump coughs, sputters, and springs to life, producing a steady stream of liquid into the plastic buckets. It gushes like a purple fountain and smells like divine blueberries as it fills the bucket. So far, it looks clean and brilliant in color, exactly what I'm looking for. I want to clarify the wine and leave the sediment behind.

With the pump working, we can relax and catch up. Certain parts of wine making are boring and tedious, so having company helps. But it has to be the right kind of company. Talking is good—up to a point—but we still need to stay focused on the work. The goal is to remove the wine from the top barrels, wash them out, and refill them with wine. It's not very complicated, but once people get to talking and sampling, mistakes can be made.

Chris owns several rental properties and always has colorful stories about drains backing up, tenants making messes, huge contractor bills to be paid, and mounds of cat hair accumulating in one tenant's apartment.

"You should consider buying property," he says. "It's a good investment."

"I'm sure it is," I say, laughing. "But I think I'll wait on that."

After we finish draining the first barrel, I use my headlamp to make sure the barrel is empty. I grab one side while Chris grabs the other. We lift it off the rack and bring it out to the driveway.

"Would you like me to wash it out?"

"Yes." I hand him the hose.

We follow the same procedure with the other barrels. Afterward, we wash up and head upstairs for dinner with my wife and daughter. I chose a menu easy to pair with the wine: steak, roasted potatoes, and a Caesar salad. I want to keep things simple and focus on the wine. Even though blind tasting can improve and sharpen your palate, as I learned while working at Betz Family Winery, I most enjoy tasting a bottle of wine with dinner and seeing how it changes in the course of a meal. Wine is meant to be part of a meal, not just something you sample at a tasting, describe poetically, or blather endlessly about, although I'm guilty on all of those counts.

I left the bottle of 2000 upright on the counter for several hours. I want the sediment to collect in the bottom so it will be clear. The bottle has no label. It's from early in my career, when I inscribed the year and varietal on the top of the cork. For comparison, I bring out a bottle of our 2017 Syrah.

Inserting the opener into the cork, I ease it out of the bottle. Old corks can disintegrate, and I don't want it to crumble and drop back into the wine. The cork cracks, but it comes out of the bottle.

I debate whether to decant it—that is, pouring it into a glass pitcher to let it breathe. I decide against it. This can sometimes diminish the taste. I toss the salad, roast the potatoes, and pan-fry the steaks—small, juicy filets mignons—for a perfect pairing, I hope. If the 2000 is off, I have the 2017 as backup.

I pour myself and Chris a glass of the 2000. Lisa and Marie don't drink.

"It smells like port," Lisa says, taking a sniff.

"It smells like wine," Marie says.

“That’s good.” I swirl it around in the glass. It’s brownish orange and remarkably translucent. The wine differs greatly from the 2017 Syrah, which is dark purple, meaty, beaty, big, and bouncy.

I sniff the wine. No vinegar smell. No nail polish scent. It smells like oranges, camphor, nuts, and glorious, transcendent fruit, where the grapes go when they reach the afterlife. I take a sip and swirl it around my mouth. It’s rich, smooth, appealing; the tannins are present but gentle. Age has improved it, not ruined it. A metaphor for our own lives?

“To your health,” Chris toasts me and digs into his steak. I take a bite of the steak and pour a glass of the 2017 for comparison. The 2017 has a beautiful core of fruit. They both enhance the steak, bringing out its meaty richness.

Remembering Robert Mondavi, I toast Chris, Lisa, and Marie. Yes, wine is a blessing, a gift, and a pure and elemental joy that puts us in touch with the earth and each other.

I taste and sip, enjoying the wine and the company. I don’t want to rush this meal. I think back on when Marie first arrived in the world and all the changes since.

“I remember when you were four,” Lisa says to Marie. “You wore the Girl Scout dress around all the time. You’d never take it off, not even to get it washed.”

“Mom, don’t embarrass me,” says Marie, who now wears a short skirt, a black mesh blouse, and black boots.

The wines evolve in the course of the meal. The 2000 remains my favorite, light and diaphanous, even after twenty years. The 2017 is dark, rich, bursting with blueberries, but rough around the edges. Age may help, but so would technique. The 2000 shows I’ve made some progress, but the 2017 demonstrates I have a way to go. Still, tasting the wines with dinner is a pleasure and the way they were made to be enjoyed. It reminds me of all the years that have passed since I started making wine some twenty-five years ago.

23

The Connoisseur

Dan McCarthy

It's a clean, well-lighted place with a tang of yeast, a hint of wood, and a long, lingering effervescent finish. McCarthy & Schiering Wine Merchant's Queen Anne shop is located near the top of Queen Anne hill, just north of downtown Seattle. Tidy wooden shelves cradle hundreds of bottles of vino from around the planet: Burgundy, Bordeaux, Champagne, Provence, Alsace, Mosel, Barolo, Barbaresco, Brunello, Rioja, Douro, Porto, Napa, Sonoma, Willamette Valley, Columbia Valley, Walla Walla, Australia, New Zealand, South Africa, Argentina. The shop has no hot pink signs, no loud promotions—just an understated elegance and devotion to one of the world's most civilized pleasures. On Saturday afternoons, a well-heeled clientele of physicians, lawyers, and tech workers crowd into the store for a free tasting of the latest offerings.

Dan McCarthy sits at a stool in the center of a large L-shaped counter, joking with customers, giving advice, and sharing samplings of the latest offerings. Dan is a gregarious, ginger-haired man in his seventies with freckled skin, lively blue eyes, and one of the best palates in Seattle. He's a true connoisseur who is focused on quality and able to cut through the hype surrounding wine.

I asked him once about buying a bottle of first-growth Bordeaux costing five hundred dollars or more. "That's a rich man's game, Nick," he said and suggested a bottle of Bordeaux's Château Pibran costing twenty-five. I was impressed and became a regular at the shop.

Having opened in 1990, the shop hosts its Saturday wine tastings with samples from around the world. It's a great education for the palate, trying Vietti Barolo from Piedmont, Italy; Trimbach Riesling from Alsace, France; and Père de Famille Cabernet from Betz Family Winery in Washington. When my kids were little and travel was difficult, the shop provided a great outlet. Each bottle represented a distilled climate—the rich Pinot Noir from Burgundy, the austere red blends of Bordeaux, the tart and sweet German Rieslings—and provided a sense of travel without leaving home.

Winemakers from around the globe visit the shop to sell their wares. Dan tastes some fifty new wines a week and keeps careful notes, which he writes up for his newsletter. He features the best of the wines at the weekly tastings. The shop is the vinous equivalent of an independent bookstore, as he promotes new wineries and established independent brands. The shop has served as an incubator for many local winemakers, including Ben Smith of Cadence and the late, great Ross Andrew Winery. In the wine business, big brands carry a great deal of clout, making it difficult for boutique houses to stand out. Dan gives them a shot.

During the week, wine sellers stop at the shop to uncork their latest offerings. That's my goal today. I bring a bottle of my Syrah for Dan to taste. I've wanted to do this for several years but have chickened out for fear of getting a bad review, keeping in mind the line from the Monty Python wine critic sketch: "This wine tastes like an Aborigine's armpit!" Finally, he suggested I bring a bottle in for a critique. I arrive at 3:00 p.m. after another seller is getting ready to leave.

Dan uncorks the bottle and takes out two glasses. He pours a splash into each, hands one to me, and takes the other. I'm nervous, wondering what he'll think. I love the wine, but it's like one of my children. I'm so close to it, it's hard for me to judge.

He swirls the wine to aerate it, takes a sip, swishes it around in his mouth as if he's gargling, and then spits it into a small bucket.

I do the same, relishing the taste and smell of the wine, which has a peppery nose, a palate of blueberries, and a lingering finish.

He nods his head appreciatively. "Where do you get your grapes?"

"Ciel du Cheval."

"You're lucky," he says. "That's a great site."

"I know," I say, wondering if he likes the wine.

"How did you ferment them?" he asks.

"We fermented them for two weeks in a tote," I say. "Erica gave us the recipe."

He nods and swirls the wine again. He's a fan of Erica Orr, our wine consultant, whose wine he sells.

"What can I improve on?" I ask.

He sniffs the wine again. "There's some oxidation in the wine. Do you top off the wine regularly?"

"Yes," I say, "but I could probably do a better job of that."

"How about your barrels?"

"I usually buy one new barrel a year," I say. "I rotate the three other barrels."

"How do you store them?"

"Mostly with wine. After bottling in the summer, I fill two with water and a sulfite and tartaric acid solution."

He smiles. "Keep it up. Keep the barrels clean. Top off the barrels."

He takes another sip and then smiles. "You've got a purple thumb, Nick," he says. "When are you going commercial?"

I laugh, thrilled he likes the wine. "I'm not sure," I say, knowing that he has promoted many local garagistes and helped them go commercial.

I'm flattered and trust his judgment, but do I want to compete with places like Opus One, Stag's Leap, Cain, Chateau Ste. Michelle, Betz, DeLille, Beaux Frères, Ken Wright, or Soter? This idea strikes me as insane. I've learned enough in my apprenticeship to know how much work and dedication it takes to succeed as a winemaker not just in the wine making but also in the viticulture, the selling, the marketing, and the distribution. And the expense! I remember the cathedral-like barrel room at Canoe Ridge Vineyard and shake my head at the cost of all those new oak wine barrels at $1,000-plus a pop. While I love to make wine, I'm not sure I want to do it as a business.

I looked into it once. I recoiled at the stacks of forms, the red tape, the state and federal taxes—all the bureaucratic hoops. Starting a nuclear waste plant seemed easier.

I hand him back the glass.

"You should consider it," he says. "Take a look at some other Syrahs for inspiration. Do you know Domaine Tempier of Provence?"

"Yes," I say. It is one of the renowned winemakers of southern France.

"They do a magnificent job with Syrah," he says. "I could arrange an appointment for you if you're going to visit."

"I'd love to do that," I say. "Let me get back to you on the dates."

I thank Dan and leave the shop. I'm pleased he likes the wine, but I'll have to think carefully before I make it my next career.

24

The Heart of Hospitality

Searching for the Spirit of Domaine Tempier

Is this the place? The old *mas* (farmhouse) stands at the end of a long gravel road, surrounded by pines, cypresses, and the deep blue dome of the Provençal sky. Grape vines radiate in all directions, their bright green leaves unfolding in the warm spring weather. The house's thick ochre walls, blue shutters, and orange tile roof give it a feeling of stability and permanence, a world away from the noise and tumult of the A50 autoroute up the valley.

My wife, Lisa, and I park our rental car in the driveway and walk along a row of tall sycamore trees, cicadas singing in the background as if in a Marcel Pagnol novel. I'm looking for Domaine Tempier, a winery epitomizing the local, organic, sustainable ideal long before it became popular in the United States. It was here that Alice Waters, who helped pioneer American organic cuisine, got her inspiration, as did countless other journalists, chefs, and winemakers who enjoyed its legendary hospitality.

But that was over thirty years ago. Would I still find the same love of great food, wine, and company that Waters described? Would I find inspiration for my own wine making during the visit?

Located among vineyards above the fishing port of Bandol, thirty miles east of Marseilles, Domaine Tempier has belonged to the Peyraud family for generations. The vineyard's proprietor is "impassioned, exuberant, dedicated to the belief that the meaning of life lies in love and friendship and that these qualities are best expressed at table," according to Richard Olney in *Lulu's Provençal Table: The Exuberant Food and Wine from the Domaine Tempier Vineyard.*

Olney's cookbook made public what many privately knew: While Lucien Peyraud was one of the finest winemakers in France, his wife, Lulu, was one of the best cooks in the country. A dynamo of a woman with sparkling eyes, dark hair, and a warm smile, Lulu welcomed an ever-expanding circle of family, friends, and visitors to her table for tapenade, brandade, guinea fowl, roast lamb, and brown bread topped with sea urchin roe. The book celebrates not only her cooking but also her infectious joie de vivre.

In a food and wine industry often driven by scores, snobbery, and faddishness, could simple pleasures of love and friendship still find expression? Can hospitality survive in the hospitality industry?

As we approach the building, I spot the Domaine Tempier sign. When I knock on the door, Annick, the receptionist, a cheerful blonde woman, welcomes us in.

"Daniel Ravier will be with you shortly," she says, leaving us to explore the tasting room. With its stone walls, wood-timbered ceilings, and track lighting, it looks like an upscale version of the old estate.

Daniel Ravier arrives and shakes hands with us. Tall, dark-haired, and jovial, he wears jeans, a pink dress shirt, and an expression of humorous bemusement. He has just finished meeting with the Peyraud family and looks as if he needs a drink.

The family hired him as the estate manager in 2000 after brothers François and Jean-Marie Peyraud retired, and the next generation declined to take over day-to-day operations. François, Jean-Marie, and their sisters remain actively involved in the business, likely making Ravier's job a challenging one.

"It's going the way I want," Ravier says, laughing. "But they could fire me in five minutes."

He opens a side door and leads the way to the winery. As we tour the grounds, he tells the story of the estate. The domaine was founded in 1834 by Lulu Tempier's family, but over the years it fell into neglect. When she and Lucien Peyraud married in 1936, they sought to rebuild its reputation. In 1945 Lucien became president of the Bandol winegrowers' association, working tirelessly to improve the estate and earn *appellation d'origine contrôlée* (controlled designation of origin) status for the region. While he managed the estate, Lulu raised the children, welcomed visitors, and became famous for her warmth, hospitality, and excellence as a cook, epitomizing the Provençal lifestyle.

"We believe in the terroir," Ravier says, gesturing to the vineyards surrounding us. "We are organic and partly biodynamic, but the terroir is the most important thing."

This terroir gives the wines—both the rosé, which comes mostly from twenty-year-old vines, and the red, which comes from forty- to fifty-year-old vines—their deep, rich, complex flavor.

Inside the winery, Ravier begins our tasting with the 2009 Domaine Tempier rosé, which he has just bottled. He pours each of us a glass. The wine displays a characteristic spicy nose with a whiff of sulfur.

"Do you add sulfur at bottling?" I ask.

"Yes, we do," he says. "It keeps the wine from going off." This step had been omitted in the past, leaving the wine vulnerable to spoilage, especially if shipped overseas. Ravier seems to have brought things up to the present without sacrificing the spirit of the estate.

Swirling the wine in the glass, I appreciate its lovely salmon color, weight (more substantial than most rosés), and exotic, spicy flavor, expressing the exuberant personality of the estate.

"It's mostly Mourvèdre and Grenache," Ravier says, "with some Cinsault and Carignan." He sniffs it, tastes it, nods approvingly, and spits it into the bucket.

Then he leads the way downstairs to the cellar, a large, cavernous room containing rows of immense wood *foudres* (casks), ideal for aging Bandol reds. These fifty-hectoliter casks impart little oak flavor, ensuring that the resulting wines will express their true character.

"We don't like woody wines," he says. "We use big vats for aging, not the smaller oak barrels."

This is the spirit of the Domaine Tempier of old, delightfully distinct from the modern international style of wine that relies on oak barrels to produce a sweet and pleasant but monotonous taste. Theirs is the approach I take with my own wine.

Ravier opens a valve on the huge oak cask and extracts a sample. "We'll start with the Cuvée *classique*," he says. "It contains mostly Mourvèdre with some Grenache, Cinsault, and Carignan."

The wine is ruby in color, with powerful fresh fruit flavors, plenty of tannins, and little trace of the oak taste that predominates in so many new-world wines.

"It's easy to do the assemblage," he says. "You look for the tears of the wine in the glass. If they go down too fast, it means you have too much alcohol. You try for balance."

He tastes the wine and then spits a long stream into the tasting bucket.

Next we move to the components of the red blend, which are sometimes bottled individually: La Tourtine, mostly Mourvèdre; La Migoua, half Mourvèdre with a mixture of Grenache, Cinsault, and Syrah; and Cabassaou, mostly Mourvèdre and aged an average of forty years. Each component displays a strong personality, worthy of being bottled on its own.

Then we taste the same wines from 2007. Dark cherry fruits emerge from behind the tannins. The wines are softer and more approachable but still make my mouth pucker.

"The wines can age from two to twenty years, depending on what you like," Ravier says. "Usually six to fifteen years is ideal, depending on the vintage. For example, the 1982 vintage is still very powerful and full of life."

After the tour, he leads us back to the reception. We thank him, and he returns to his office. We linger in the reception room, not wanting the tour to end. As much as I've enjoyed it, I feel a little disappointed. What about the legendary Domaine Tempier hospitality? I know I can't expect the extended table underneath a canopy of vines with sea urchin roe spread on a cracker by Lulu, but everything I'd read about the place created such expectations.

"The Peyraud's family's example has been helping us find our balance at Chez Panisse for years," says Alice Waters in the foreword to *Lulu's Provençal Table*. "Like them, we try to live close to the earth and treat it with respect; always look first to the garden and the vineyard for inspiration; rejoice in our families and friends; and let the food and wine speak for themselves at the table."

During the tour, I've caught glimpses of the Domaine Tempier of old, especially in the wine making, but still I crave something more. I decide to buy three bottles of red wine as keepsakes. The credit card machine is slow to process my payment, on strike like so many things in France.

An old wooden door creaks open. Like an apparition, out steps ninety-three-year-old Lulu Peyraud wearing a floral print dress. I try not to stare; I didn't realize she was still alive. Her hair has gone gray, and she walks with a cane, but she smiles warmly.

"I've enjoyed your recipes," I tell her in French.

"Merci." Her blues eyes sparkle in welcome. "Bienvenue et bonne degustation." After welcoming us and wishing us a good tasting, she returns to the family's quarters behind the facade of the modern reception room. The spirit of Domaine Tempier lives!

I stumble out the door into the bright Provençal sunshine, grateful to have caught a glimpse of Lulu and the legendary Domaine Tempier hospitality. I carefully clutch the three bottles of precious Domaine Tempier wine, which will serve as inspiration for our big summer bottling party back home in the United States.

25

Bottling Party

It's the highlight of the wine year. After the drama of the crush, fermentation, topping, and racking, then comes the bottling party, when the wine leaves the barrels for the bottles and the time, effort, and expense come to fruition. It's incredibly satisfying, a kind of crescendo requiring coordination of many moving parts. It has to be well organized, or the whole thing could flop.

The planning starts months in advance. Like the conductor of the orchestra, I have to round up the key players—Tom, Bruce, Chris, John, Gregory, Abbey Bennett, and Steve Rottler—for a date we can all agree on. Once that's settled, I send out the announcement to our Les Copains email list:

> You are cordially invited to participate in our annual bottling party starting at 2 p.m. at my place.
>
> We'll bottle our meaty, beaty, big, and bouncy Syrah from Ciel du Cheval, one of the top vineyards in Washington State. Afterward, we'll enjoy a potluck feast around 7 p.m. that will include wines from my trip to Tuscany, Provence, and Rioja.
>
> Please bring a side dish that doesn't require a lot of preparation and elbow grease, and a hearty appetite and thirst. Last names beginning with *A* to *G*, bring a main course (I'll be smoking some ribs); *H* to *N*, bring side dish; all the rest, bring dessert, appetizer, or bread and cheese. No high heels or fancy clothes or lizard skin cowboy boots during bottling, which can be quite messy, though such attire can be appropriate at the dinner.
>
> Everyone who helps will get at least one free bottle of wine, and there will be lots of opportunities for quality control—i.e., tasting. Those coming from afar will get ONE free bottle per time zone

traveled. If you haven't done so already, please let me know if you can make it!

Once the RSVPs trickle in, I'll know if we'll have enough help.

Next, I order wine bottles and corks from one of the big glass companies. I have to do this far enough ahead so they arrive prior to the party. It's a disaster if they arrive late, which has happened, so I work hard to prevent it.

We produce some eighty cases, or around a thousand bottles, which is a sizable amount for amateurs but tiny compared to the boutique wineries where I've worked. Still, the process is similar, and I try to apply all I've learned in my apprenticeship: ordering primo corks and bottles, insisting on quality control throughout the operation.

We're so small we barely qualify to order from the commercial glass companies that operate according to economies of scale. I think they sell to us in the hope that we'll go commercial and become bigger clients. I don't tell them that's not going to happen as I enjoy interacting with the salesmen. Though not as "romantic" as growing grapes or making wine, purchasing corks and bottles is a critical part of the process.

Next, I arrange with the company that it will send a truck small enough to navigate the narrow streets of north Queen Anne hill but large enough to carry the pallet of wine bottles. When the truck arrives, I help the driver lower the pallet on the truck's lift and guide it into position on the side of the driveway. I cover it with a blue tarp to keep it ready for bottling day.

On bottling day, I wrestle the stainless steel bottling machine out of the attic and set it up on a table in the driveway. The machine has a simple, timeless design with a tank on top that has a float to regulate the flow into five nozzles at the front. The nozzles can be adjusted to automatically fill the bottles right above the shoulder, ensuring there's little air left before corking.

Tom arrives around two o'clock, bringing hoses and buckets. We put a wand and siphon hose in the first barrel and attach it to the pump. Then we run another section of hose along the ceiling of my garage, up the side of the house, and into a fifty-gallon white plastic bucket. We set up a siphon from the bucket into the bottling machine.

"Where's the wrench?" Tom asks, working on the pump.

I fetch one for him.

Volunteers start arriving. We need at least ten people to operate at full capacity: one person to oversee the pump, another to watch the bucket and siphon, another to operate the bottling machine, two people for the corker, one person for labels, another person to fill the empty cases with the bottles, and others to keep track of the drinks and food for the workers.

As at Betz and DeLille, we assign our veterans—especially Bruce McLachlin, who is a key member of the team and our Mr. Fix It—to instruct the newbies. Tall, good-natured, and a mechanical genius, he keeps everything working properly.

Tom starts the pump. First water and then wine flows through the hoses and up into the big plastic bucket.

Bruce watches the bucket above the bottling machine, waiting until the water clears the line. Then he shoves the hose into the bucket. The dark purple liquid splashes against the side. "We got wine!" he says.

"Tell me when it's full!" Tom yells from inside the garage.

The pump strains. The dark, glistening liquid races through the tube and fills the bucket.

"When!" Bruce yells.

Tom shuts off the pump.

Next, Tom and Bruce set up the siphon from the bucket down to the bottling machine. Once it's running, wine flows from the bucket by gravity into the bottling machine. Gregory makes sure each of the filling nozzles in the machine is working. Wearing a wide-brimmed sun hat, he brings gusto to the job, working the bottling machine like a synthesizer.

As he fills the first bottle, the machine automatically directs the wine into the bottle but allows too much air. He adjusts it so the wine fills right above the shoulder of the bottle. He hands the bottle to the corkers, who place the cork and bottle into the corking machine.

At first the crew runs haphazardly, learning the ropes. Abbey takes the bottles from Gregory, wipes them off, and hands them to the corker, Steve. Another volunteer serves as a quality checker, making sure the bottle is filled just above the shoulder—no higher and no lower.

The corking station requires two people: one to place a cork and bottle into the machine and the other to pull down the handle that forces a metal flange to squish and insert the cork into the bottle.

Once the volunteers get the hang of it, the bottling line starts cranking. Everyone joins in to make it happen.

"I need more corks," Chris Olsen says.

I find more.

"Can I get a clamp for the siphon hose?" Bruce asks.

I find one and toss it up to him.

When the questions die down, I dip a plastic cup into the wine from the bottling machine. I swirl and sip it. It tastes fantastic. Now we have to keep it that way.

"Where are the plastic cups?" someone yells.

I track those down.

It's my job to keep the volunteers happy, getting them plastic cups, bottled water, paper towels, and whatever else they need.

"We need help with the labels," says Theresa Jurotich, a thin, trim triathlete with short blonde hair who is in charge of the labels.

I find a volunteer to help her.

Returning to the bottling line, I check the last case that has been corked. I hold up a bottle and see that there's too much air between the cork and the shoulder of the bottle. All the bottles in this case will need to be refilled and recorked, taking precious time. The other cases look fine.

"Quality control!" I yell, holding up the bottles.

I hand over a corkscrew to a volunteer who removes the corks. The team fills up the bottles and recorks them. No more tippling until the bottling crew gets it right.

As everyone gets in the groove, we work faster. We need to finish by seven o'clock for dinner.

I hear the pump straining. I run back into the garage to check on it. Wine is spurting out of the pump and pooling on the floor. "Tom, we need to fix the pump!"

Tom shuts it off and goes to work on it. The entire bottling line has to shut down. "Tell everybody to take a break," he says.

"Will do," I say, worried we won't finish on time.

He disassembles it. "We need a new ring. Do you have one?" The rubber ring serves as a seal, keeping the liquid from dripping out of the pump. The old ring broke.

I rummage around in the cabinets. I try not to think about what will happen if I can't find it. The whole operation could shut down because of a tiny rubber ring. Then I see a small box. Bingo! I find a replacement ring for the pump.

"Here you go," I say as I hand it to Tom, who replaces the ring. He restarts the pump. The leak is sealed.

"Let's get cranking," he says.

"Coffee break's over," I say, getting the bottling line back together.

It's late in the afternoon. Tom pumps the last of the wine out of the barrel and into the bucket. Bruce watches as the siphon sips the last of the liquid. The operation reaches a crescendo as the volunteers start the countdown for the last bottles filled.

"Five!"

"Four!"

"Three!"

"Two!"

"One!"

With that, the bottling is finished. We high-five each other and take a sip of wine. Now we have to clean the machinery and get ready for dinner. The bottling machine needs to be washed; Tom and Gregory work on that. The barrels needed to be hosed out, treated with sulfites, and refilled with water. That's my job.

I've learned that wine making is less about romance than washing, washing, and washing. Will the washing ever end? To paraphrase Lady Macbeth, "What, will these barrels never be clean?"

The Summer Party

I head up to the deck and inhale the delicious smell of smoked ribs. I put them in the smoker five hours ago, and now they smell divine. I head inside to see how the dinner is progressing.

"How many people are staying for dinner?" Lisa asks.

"I'll check," I say, rushing back outside.

"I wish people would RSVP," she says.

"I know," I say, "but it always seems to work out."

Petite, red-haired, and introverted, Lisa prefers order to chaos. As a part-time extrovert, I relish a little chaos, especially now that the wine is in the bottles.

After boiling the pasta, Lisa adds basil, tomatoes, onions, and prosciutto. My sister Anne Weglin, an enthusiastic cook whom we refer to as Martha Stewart, cuts the onions. My sister Cathy Rouleau, blonde haired and blue eyed with an appreciation of the absurd, cores the tomatoes. They make several different kinds of pasta for a sumptuous meal for the multitudes.

"Does anybody want steak?" Tom breezes past carrying a heaping platter of meat.

"Why does he want steak if we have so much pasta?" Lisa asks.

"It doesn't matter," I say. "He can cook it outside on the grill."

Lisa, my sisters, and others have set up long tables in the backyard with folding chairs, bright green tablecloths, napkins, and cutlery. They invite people into the living room to help themselves to pasta, steak, ribs, green salad, salmon, and fruit tarts.

Finally, I relax. Everyone gathers in our backyard at the tables underneath the pear tree. The air is warm and dry, plenty of light is in the sky, and a crescent moon is pinned to the horizon. The women wear summer dresses in a variety of pastels and patterns. The men wear Hawaiian shirts and sandals, as if they're going to a Jimmy Buffett concert. Also present are students from my travel writing classes in Europe, friends from years past, and fellow parishioners from St. Anne Catholic Church on Queen Anne hill. The summer bottling party is an excuse to get together and enjoy each other's company, highlighting the communal aspect of wine and food.

Everyone digs into the food. The wine flows freely. A glow suffuses our faces. Is it the wine or happiness? Both.

I reflect on how much I've learned in the apprenticeship, not just about grapes, vines, and terroir, but about work, vocation, character, and love for the craft of wine making as well.

I raise a glass. "Here's to all of you," I say. "There's a reason we named it Les Copains: We couldn't do it without you." And then I add what I've learned during my apprenticeship: "And I wouldn't want to."

I've loved all the places I've visited: Opus One, Stag's Leap, Cain, Chateau Ste. Michelle, Betz, DeLille, Beaux Frères, Ken Wright, and Soter. But I don't want to have to worry about wine scores, sales meetings, production schedules. That would take the joy out of wine making for me.

The scene resembles a painting by the Dutch artist Pieter Bruegel: Kids devour the ribs and smear BBQ sauce all over themselves; a couple kisses over a glass of Les Copains; Tom and his wife, Jen, erupt in raucous laughter; our border collie, Stella, lurks underneath the table and inhales a scrap of meat from Bruce. It's a timeless moment, combining the ephemeral and the eternal. Here, now, always.

BIBLIOGRAPHIC ESSAY

Most of the research for *Crush* involved first-person participation and observation of wine making; however, I also conducted interviews and researched the history of American wine and winemakers. My definition of "terroir" includes not only site, soil, and weather but also the individuals who plant the grapes and make the wine. Their personalities shape the industry and provide inspiration for later generations of winemakers and wine aficionados like me.

Selected Sources

Blecha, Peter. "Holmes, Jim (b. 1936)." Historylink.org, June 30, 2020. https://www.historylink.org/file/21045.

Griffin, Rob. Kiona website, "1978." https://kionawine.com/the-story-and-people.

Irvine, Ronald, and Walter J. Clore. *The Wine Project: Washington State's Winemaking History*. Vashon Island WA: Sketch Publications, 1997.

Mondavi, Robert. *Harvests of Joy: How the Good Life Became Great Business*. New York: Harvest, 1999.

Olney, Richard. *Lulu's Provençal Table: The Exuberant Food and Wine from the Domaine Tempier Vineyard*. Berkeley CA: Ten Speed Press, 2002.

Steury, Tim. "Walter Clore: A Wine Visionary." *Washington State Magazine*, Washington State University (Fall 2003). https://magazine.wsu.edu/2009/10/09/walter-clore-a-wine-visionary/.

Taber, George M. *Judgment of Paris: California vs. France and the Historic 1976 Paris Tasting That Revolutionized Wine*. New York: Scribner, 2005.

INDEX